W9-BPK-431

The NEW ENCYCLOPEDIA *of* SOUTHERN CULTURE

VOLUME 22 : SCIENCE & MEDICINE

Volumes to appear in
The New Encyclopedia of Southern Culture
are:

Agriculture and Industry	*Law and Politics*
Art and Architecture	*Literature*
Education	*Media*
Environment	*Music*
Ethnicity	*Myth, Manners, and Memory*
Folk Art	*Race*
Folklife	*Religion*
Foodways	*Science and Medicine*
Gender	*Social Class*
Geography	*Sports and Recreation*
History	*Urbanization*
Language	*Violence*

The NEW

ENCYCLOPEDIA *of* SOUTHERN CULTURE

CHARLES REAGAN WILSON General Editor

JAMES G. THOMAS JR. Managing Editor

ANN J. ABADIE Associate Editor

VOLUME 22

Science & Medicine

JAMES G. THOMAS JR. &
CHARLES REAGAN WILSON
Volume Editors

Sponsored by
THE CENTER FOR THE STUDY OF SOUTHERN CULTURE
at the University of Mississippi
THE UNIVERSITY OF NORTH CAROLINA PRESS
Chapel Hill

This book was published with the
assistance of the Anniversary Endowment Fund
of the University of North Carolina Press.

Designed by Richard Hendel
Set in Minion types by Tseng Information Systems, Inc.
Manufactured in the United States of America
The paper in this book meets the guidelines for permanence and durability of the Committee on Production Guidelines for Book Longevity of the Council on Library Resources.
The University of North Carolina Press has been a member of the Green Press Initiative since 2003.
Library of Congress Cataloging-in-Publication Data
Science and medicine / James G. Thomas Jr. and
Charles Reagan Wilson, volume editors
p. cm. — (The new encyclopedia of Southern culture ; v. 22)
"Sponsored by The Center for the Study of Southern Culture at the University of Mississippi."
Includes bibliographical references and index.
ISBN 978-0-8078-3719-1 (cloth : alk. paper) —
ISBN 978-0-8078-3720-7 (pbk.: alk. paper)
1. Science—Southern States—Encyclopedias. 2. Medicine—Southern States—Encyclopedias. 3. Southern States—Social conditions—Encyclopedias. I. Thomas, James G. (James George). II. Wilson, Charles Reagan. III. University of Mississippi. Center for the Study of Southern Culture. IV. Series.
F209 .N47 2006 vol. 22
[Q127.U6]
975.003 s—dc22
2012655018
The *Encyclopedia of Southern Culture*, sponsored by the Center for the Study of Southern Culture at the University of Mississippi, was published by the University of North Carolina Press in 1989.
cloth 16 15 14 13 12 5 4 3 2 1
paper 16 15 14 13 12 5 4 3 2 1

Tell about the South. What's it like there.

What do they do there. Why do they live there.

Why do they live at all.

WILLIAM FAULKNER

Absalom, Absalom!

CONTENTS

GENERAL INTRODUCTION

In 1989 years of planning and hard work came to fruition when the University of North Carolina Press joined the Center for the Study of Southern Culture at the University of Mississippi to publish the *Encyclopedia of Southern Culture*. While all those involved in writing, reviewing, editing, and producing the volume believed it would be received as a vital contribution to our understanding of the American South, no one could have anticipated fully the widespread acclaim it would receive from reviewers and other commentators. But the *Encyclopedia* was indeed celebrated, not only by scholars but also by popular audiences with a deep, abiding interest in the region. At a time when some people talked of the "vanishing South," the book helped remind a national audience that the region was alive and well, and it has continued to shape national perceptions of the South through the work of its many users—journalists, scholars, teachers, students, and general readers.

As the introduction to the *Encyclopedia* noted, its conceptualization and organization reflected a cultural approach to the South. It highlighted such issues as the core zones and margins of southern culture, the boundaries where "the South" overlapped with other cultures, the role of history in contemporary culture, and the centrality of regional consciousness, symbolism, and mythology. By 1989 scholars had moved beyond the idea of cultures as real, tangible entities, viewing them instead as abstractions. The *Encyclopedia*'s editors and contributors thus included a full range of social indicators, trait groupings, literary concepts, and historical evidence typically used in regional studies, carefully working to address the distinctive and characteristic traits that made the American South a particular place. The introduction to the *Encyclopedia* concluded that the fundamental uniqueness of southern culture was reflected in the volume's composite portrait of the South. We asked contributors to consider aspects that were unique to the region but also those that suggested its internal diversity. The volume was not a reference book of southern history, which explained something of the design of entries. There were fewer essays on colonial and antebellum history than on the postbellum and modern periods, befitting our conception of the volume as one trying not only to chart the cultural landscape of the South but also to illuminate the contemporary era.

When C. Vann Woodward reviewed the *Encyclopedia* in the *New York Review of Books*, he concluded his review by noting "the continued liveliness of interest in the South and its seeming inexhaustibility as a field of study." Research on the South, he wrote, furnishes "proof of the value of the *Encyclopedia* as a scholarly undertaking as well as suggesting future needs for revision or supplement to keep up with ongoing scholarship." The two decades since the publication of the *Encyclopedia of Southern Culture* have certainly suggested that Woodward was correct. The American South has undergone significant changes that make for a different context for the study of the region. The South has undergone social, economic, political, intellectual, and literary transformations, creating the need for a new edition of the *Encyclopedia* that will remain relevant to a changing region. Globalization has become a major issue, seen in the South through the appearance of Japanese automobile factories, Hispanic workers who have immigrated from Latin America or Cuba, and a new prominence for Asian and Middle Eastern religions that were hardly present in the 1980s South. The African American return migration to the South, which started in the 1970s, dramatically increased in the 1990s, as countless books simultaneously appeared asserting powerfully the claims of African Americans as formative influences on southern culture. Politically, southerners from both parties have played crucial leadership roles in national politics, and the Republican Party has dominated a near-solid South in national elections. Meanwhile, new forms of music, like hip-hop, have emerged with distinct southern expressions, and the term "dirty South" has taken on new musical meanings not thought of in 1989. New genres of writing by creative southerners, such as gay and lesbian literature and "white trash" writing, extend the southern literary tradition.

Meanwhile, as Woodward foresaw, scholars have continued their engagement with the history and culture of the South since the publication of the *Encyclopedia*, raising new scholarly issues and opening new areas of study. Historians have moved beyond their earlier preoccupation with social history to write new cultural history as well. They have used the categories of race, social class, and gender to illuminate the diversity of the South, rather than a unified "mind of the South." Previously underexplored areas within the field of southern historical studies, such as the colonial era, are now seen as formative periods of the region's character, with the South's positioning within a larger Atlantic world a productive new area of study. Cultural memory has become a major topic in the exploration of how the social construction of "the South" benefited some social groups and exploited others. Scholars in many disciplines have made the southern identity a major topic, and they have used a variety of methodologies to suggest what that identity has meant to different social groups. Literary critics have adapted cultural theories to the South and have raised the issue of postsouthern literature to a major category of concern as well as exploring the links between the literature of the American South and that of the Caribbean. Anthropologists have used different theoretical formulations from literary critics, providing models for their fieldwork in southern communities. In the past 30 years anthropologists have set increasing numbers of their ethnographic studies in the South, with many of them now exploring topics specifically linked to southern cultural issues. Scholars now place the Native American story, from prehistory to the contemporary era, as a central part of southern history. Comparative and interdisciplinary approaches to the South have encouraged scholars to look at such issues as the borders and boundaries of the South, specific places and spaces with distinct identities within the American South, and the global and transnational Souths, linking the American South with many formerly colonial societies around the world.

The first edition of the *Encyclopedia of Southern Culture* anticipated many of these approaches and indeed stimulated the growth of Southern Studies as a distinct interdisciplinary field. The Center for the Study of Southern Culture has worked for more than three decades to encourage research and teaching about the American South. Its academic programs have produced graduates who have gone on to write interdisciplinary studies of the South, while others have staffed the cultural institutions of the region and in turn encouraged those institutions to document and present the South's culture to broad public audiences. The center's conferences and publications have continued its long tradition of promoting understanding of the history, literature, and music of the South, with new initiatives focused on southern foodways, the future of the South, and the global Souths, expressing the center's mission to bring the best current scholarship to broad public audiences. Its documentary studies projects build oral and visual archives, and the New Directions in Southern Studies book series, published by the University of North Carolina Press, offers an important venue for innovative scholarship.

Since the *Encyclopedia of Southern Culture* appeared, the field of Southern Studies has dramatically developed, with an extensive network now of academic and research institutions whose projects focus specifically on the interdisciplinary study of the South. The Center for the Study of the American South at the University of North Carolina at Chapel Hill, led by Director Jocelyn Neal and Associate Director and *Encyclopedia* coeditor William Ferris, publishes the lively journal *Southern Cultures* and is now at the organizational center of many other Southern Studies projects. The Institute for Southern Studies at the University of South Carolina, the Southern Intellectual History Circle, the Society for the Study of Southern Literature, the Southern Studies Forum of the European American Studies Association, Emory University's SouthernSpaces.org, and the South Atlantic Humanities Center (at the Virginia Foundation for the Humanities, the University of Virginia, and Virginia Polytechnic Institute and State University) express the recent expansion of interest in regional study.

Observers of the American South have had much to absorb, given the rapid pace of recent change. The institutional framework for studying the South is broader and deeper than ever, yet the relationship between the older verities of regional study and new realities remains unclear. Given the extent of changes in the American South and in Southern Studies since the publication of the *Encyclopedia of Southern Culture*, the need for a new edition of that work is clear. Therefore, the Center for the Study of Southern Culture has once again joined the University of North Carolina Press to produce *The New Encyclopedia of Southern Culture*. As readers of the original edition will quickly see, *The New Encyclopedia* follows many of the scholarly principles and editorial conventions established in the original, but with one key difference; rather than being published in a single hardback volume, *The New Encyclopedia* is presented in a series of shorter individual volumes that build on the 24 original subject categories used in the *Encyclopedia* and adapt them to new scholarly developments. Some earlier *Encyclopedia* categories have been reconceptualized in light of new academic interests. For example, the subject section originally titled "Women's Life" is reconceived as a new volume, *Gender*, and the original "Black Life" section is more broadly interpreted as a volume on race. These changes reflect new analytical concerns that place the study of women and blacks in broader cultural systems, reflecting the emergence of, among other topics, the study of male culture and of whiteness. Both volumes draw as well from the rich recent scholarship on women's life and black life. In addition, topics with some thematic coherence are combined in a volume, such as *Law and Politics* and *Agriculture and Industry*. One new topic, *Foodways*, is the basis of a separate volume, reflecting its new prominence in the interdisciplinary study of southern culture.

Numerous individual topical volumes together make up *The New Encyclopedia of Southern Culture* and extend the reach of the reference work to wider audiences. This approach should enhance the use of the *Encyclopedia* in academic courses and is intended to be convenient for readers with more focused interests within the larger context of southern culture. Readers will have handy access to one-volume, authoritative, and comprehensive scholarly treatments of the major areas of southern culture.

We have been fortunate that, in nearly all cases, subject consultants who offered crucial direction in shaping the topical sections for the original edition have agreed to join us in this new endeavor as volume editors. When new volume editors have been added, we have again looked for respected figures who can provide not only their own expertise but also strong networks of scholars to help develop relevant lists of topics and to serve as contributors in their areas. The reputations of all our volume editors as leading scholars in their areas encouraged the contributions of other scholars and added to *The New Encyclopedia*'s authority as a reference work.

The New Encyclopedia of Southern Culture builds on the strengths of articles in the original edition in several ways. For many existing articles, original authors agreed to update their contributions with new interpretations and theoretical perspectives, current statistics, new bibliographies, or simple factual developments that needed to be included. If the original contributor was unable to update an article, the editorial staff added new material or sent it to another scholar for assessment. In some cases, the general editor and volume editors selected a new contributor if an article seemed particularly dated and new work indicated the need for a fresh perspective. And importantly, where new developments have warranted treatment of topics not addressed in the original edition, volume editors have commissioned entirely new essays and articles that are published here for the first time.

The American South embodies a powerful historical and mythical presence, both a complex environmental and geographic landscape and a place of the imagination. Changes in the region's contemporary socioeconomic realities and new developments in scholarship have been incorporated in the conceptualization and approach of *The New Encyclopedia of Southern Culture*. Anthropologist Clifford Geertz has spoken of culture as context, and this encyclopedia looks at the American South as a complex place that has served as the context for cultural expression. This volume provides information and perspective on the diversity of cultures in a geographic and imaginative place with a long history and distinctive character.

The *Encyclopedia of Southern Culture* was produced through major grants from the Program for Research Tools and Reference Works of the National Endowment for the Humanities, the Ford Foundation, the Atlantic-Richfield Foundation, and the Mary Doyle Trust. We are grateful as well to the College of Liberal Arts at the University of Mississippi for support and to the individual donors to the Center for the Study of Southern Culture who have directly or indirectly supported work on *The New Encyclopedia of Southern Culture*. We thank the volume editors for their ideas in reimagining their subjects and the contributors of articles for their work in extending the usefulness of the book in new ways. We acknowledge the support and contributions of the faculty and staff at the Center for the Study of Southern Culture. Finally, we want especially to honor the work of William Ferris and Mary Hart on the *Encyclopedia of Southern Culture*. Bill, the founding director of the Center for the Study of Southern Culture, was coeditor, and his good work recruiting authors, editing text, selecting images, and publicizing the volume among a wide network of people was, of course, invaluable. Despite the many changes in the new encyclopedia, Bill's influence remains. Mary "Sue" Hart was also an invaluable member of the original encyclopedia team, bringing the careful and precise eye of the librarian, and an iconoclastic spirit, to our work.

INTRODUCTION

When a reader thinks of "southern culture," images of research scientists and physicians probably do not immediately come to mind. Representations of the region often go back to 19th-century myths, icons, and social types, and science and medicine were not at the center of early imaginings of the South. Those subjects are, nonetheless, important ones for a full understanding of the region's internal dynamics and positioning in the nation. The climate nurtured certain kinds of distinctive diseases, from epidemic diseases like yellow fever to debilitating illnesses like hookworm and pellagra, the latter of which fed southern stereotypes of lazy poor white trash. White southerners promoted the belief in "states' rights medicine" as part of southern nationalism, and medical experimentation on African American bodies from antebellum gynecological studies to the 20th-century Tuskegee syphilis study expressed the worst racial underpinnings to ideas of distinctive southern medicine. The South was home to Enlightenment-influenced scientific rationalists but was later a center of national struggles between science and religion, as seen in the 1925 Scopes Trial in Tennessee. Lack of financial resources for a hundred years after the Civil War made it difficult to fund the region's universities at levels sufficient to be national leaders in scientific research. In more recent times, increased prosperity has promoted the rise in the South's scientific research profile, with industry and state governments supporting such efforts for economic development. An organization like the Centers for Disease Control and Prevention can now become a global leader in health matters. The South, nonetheless, remains haunted in health matters by specters of the region's past—poverty and racial discrepancies above all. Despite dramatic health-care improvements overall, southerners today are still prone to greater susceptibility than people in other parts of the country to certain health problems, such as high rates of obesity, HIV/AIDS, hypertension, and diabetes. As in the past, health problems affect the South's poor people disproportionately, and African Americans suffer debilitating and life-threatening conditions from socioenvironmental conditions.

The *Science and Medicine* volume of *The New Encyclopedia of Southern Culture* has dramatically expanded from the topic's treatment in the earlier edition of the encyclopedia. The overview essay charts the historical centrality of disease to southern distinctiveness and the contributions and limitations of science in the region, pointing to recent advances on both fronts and assessing the contemporary context for health and research. Thirty-eight thematic articles and 45 topical/biographical entries explore such major topics as scientific agriculture, aerospace research, public health, mental health, rural and urban health, black health, worker health, alcoholism and drug use, and maternal and child health. The editors have revealed the diversity of science and medicine in the region through articles on American Indian medicine, gender and health, women healers, folk medicine, and African American physicians. The institutional structure of southern science and medicine is outlined through such articles as Research Triangle Park, Meharry Medical College, St. Jude Children's Research Hospital, and Whitfield (Mississippi State Hospital). The diseases of the region are covered with articles ranging from malaria to hookworm to the influenza epidemic of 1918 to leprosy. The achievements of scientific and medical innovators are acknowledged with articles on such figures as heart surgeon Michael DeBakey, medical researcher Walter Reed, and chemical scientist Charles Holmes Herty. The Civil War and the civil rights movement—ever central to ponderings of the South—are shown to have medical dimensions through entries on Civil War medicine and the Medical Committee for Human Rights.

This volume opens up still-underexplored topics in southern culture but also reveals the recent work that has produced increased understanding of how research and health issues have been enduring ones in the region. Contemporary contexts show how much has changed in the region's experience and yet reveal the remaining legacies of social and environmental conditions that continue to affect the life of southerners.

The NEW ENCYCLOPEDIA *of* SOUTHERN CULTURE

VOLUME 22 : SCIENCE & MEDICINE

SCIENCE AND MEDICINE

Although little noticed by the myriad students of the South, science and medicine have been important and instructive components of southern culture. On one level, they have contributed to social progress. On another, they have been barometers for gauging intellectual life. The South's experience in these areas also sheds valuable light on the question of southern distinctiveness, providing additional support for the contention that regional separateness has had a retarding effect on cultural development.

Colonial South. An interest in science was part of the cultural baggage that the first colonists brought to the South. This interest was fed and intensified by the seemingly insatiable curiosity of Europeans regarding the natural life and products of the New World. Consequently, from the earliest days of colonization the pursuit of science was a prominent feature of southern life.

Throughout the colonial period and into the 19th century, science was generally divided into two broad categories—natural philosophy (the physical sciences) and natural history (the natural sciences). The former was concerned largely with the verification of existing scientific principles and the latter with the observation, collection, and classification of the phenomena of the natural world. Because of the physical and intellectual limitations of their frontier setting, colonial Americans were ill prepared to do much in natural philosophy, but they were ideally situated to excel in natural history. Their research in natural history set a pattern of activity that dominated American science for three centuries.

Motivated by the irresistible appeal of the lush, often exotic, natural world that surrounded them as well as by requests for assistance from English and European students of nature eager for New World botanical, zoological, and mineral specimens for their research and personal collections, hundreds of early southerners became actively involved in natural history. Most of them served as field collectors for Europeans. By the late colonial period, a few became highly competent scientists and, as respected members of the international circle of natural historians, made contributions to scientific advancement. The most important figures in southern natural history in the 17th century were John Clayton I and John Banister and, in the 18th, Mark Catesby,

John Clayton II, John Mitchell, and Alexander Garden. Garden, a Charleston physician, was perhaps the most accomplished and best known of the group.

The activities and accomplishments of the early South's natural historians were highly significant: they made the region, along with the middle colonies, the colonial leader in the study of the American natural world; they played an indispensable role in filling in the New World book of nature and thus contributed to the advancement of Western science; and they helped lay the foundation for American science.

The medical story of the early South was not nearly so bright or promising. It was in fact tragic. Disease and death were constant companions of colonists everywhere, but especially in the South. Here, health hazards, ranging from endemic "ague" (chills and fever) and "flux" (dysentery) to epidemic outbreaks of smallpox and yellow fever, were at their worst. It is easy to see why the southern colonists were less healthy. "Seasoning," or becoming acclimated to the region's semitropical climate, was the source of extreme morbidity and mortality. Moreover, because of an environment that encouraged insect life, a general disregard for the draining of swamps, and the steady influx of black carriers with the rise of slavery, malaria—early America's most dangerous endemic disorder—tightened its hold on the South in the 18th century as it began to disappear from New England. Southern forms of the disease were more debilitating and deadly than those that prevailed elsewhere in the colonies. Finally, the medical reforms of the late colonial period that improved health—more and better-trained physicians, therapeutic advancements (e.g., the use of variolation to prevent smallpox and cinchona bark to control malaria), and the gradual appearance of regional medical institutions—such as schools, societies, and licensing—made less headway in the South than elsewhere.

As the health picture of the South suggests, the American colonies exhibited regional distinctions quite early because of the diversity of colonizing experiences and New World conditions. But while New England and the middle colonies became recognizable colonial divisions, the southern colonies showed the greatest cultural diversity. Indeed, during the century and a half between the settlement of Jamestown and the outbreak of the Revolution, the seeds of southern distinctiveness were planted, and the first shoots sent up. Such regional influences encouraged a separate southern identity. The colonial South's first European settlers transplanted the social model of the English country gentleman to the New World and tried to follow it. They found favorable climatic and geographic conditions and established a plantation economy based on slavery and a staple-crop system.

Although sectional identity had little immediate meaning for the principal areas of regional life, the factors that underpinned it boded ill in the cases of science and medicine. Agrarianism and the plantation system, for example, fostered ruralism and a sparse pattern of settlement that discouraged and retarded urbanization, with its greater opportunities for intellectual contact and its nurturing environment for societies, journals, and other institutions for the promotion of science. The poor health of the South was in large measure attributable to the social consequences of the region's unswerving devotion to a way of life based on slavery and the plantation economy.

Old South. Southern colonists on the eve of the Revolution were no more devoted to sectionalism than those in New England and the middle colonies. In fact, after the break with England, they were among the most strident cultural nationalists and celebrated America's special destiny. Independence and nationhood, however, provided the impetus for the transformation of the embryonic South into the sectional South. This unintended, and largely unconscious, historical process was the result of growing inconsistencies between the southern way of life and emerging national patterns that became increasingly obvious after independence. None was more glaring than the South's slave-based economic system and its underlying racism, which stood in contradiction to the philosophy of the Revolution and the idealism of the early Republic. Forced to choose, white southerners rejected freedom and equality for slavery and racism. Such unsettling experiences led, by the end of the 18th century, to the emergence of a southern sectional consciousness—the First South. After 1820 a sense of grievance and feelings of defensiveness united white southerners as never before and pushed them further out of the national mainstream. This was the Old South, the supreme expression of southern distinctiveness. Science and medicine, like all of southern life, bore the imprint of the South's sectional philosophy.

Reflecting the cultural nationalism of the era, science in the early national period was characterized by the establishment and shaping of institutions and attitudes aimed at ending America's intellectual subservience to Europe. Among the achievements of the period were the establishment of new schools and the improvement of existing ones, the founding of scientific societies and journals, and the building of museums and herbaria. As the result of these steps, the United States by 1830 had become a junior partner with Europe in science and had started down the path that would lead to eventual leadership in the scientific world. The South's leaders of science supported and contributed to the drive for national scientific independence. In fact, Thomas Jefferson, the region's best-known scientist of the early national period, was crucial to the quest for a first-rate American science. Although Jefferson was not a great scientist, his influence permeated the pursuit of science nationwide, and he was a tower of strength to all interested in science.

America's striving for scientific respectability coincided with the maturing of Western science. Indeed, the 19th century was a golden age for science. During this century, science came of age and established its utility for social progress. The result was a veritable cult of science that affected every aspect of life. The United States, while overshadowed by the scientific leaders in Europe, was actively involved in the modernization of science.

Between 1820 and 1860, the four decades that are generally associated with the Old South, all parts of the country did not participate equally in the advancement of American science: the Northeast was the clear leader, the West contributed the least, and the South occupied an intermediate position. After performing splendidly in the colonial period, the South fell behind the Northeast in science after the Revolution. The South's comparative lag in science was seen in a variety of ways, including the production of fewer scientists than the northern and middle states, a slower pace of institutional development for the support of science, a lower level of scientific activity, and a less progressive attitude toward science.

Reasons for the region's declining national position ranged from agrarianism to the capitulation of the South to evangelical Protestantism and the defensiveness that accompanied mounting sectional tensions. These forces retarded the growth of institutions for the pursuit of science and created a climate of opinion that inhibited the free inquiry crucial to scientific advancement.

For all its problems, limitations, and comparative lag, science occupied a prominent place in antebellum southern culture. Natural history continued to be the dominant type of scientific activity. Although leadership in this area had shifted to the North by the advent of the Old South, southern contributions to the advancement of natural history were extensive and important. Students of the natural world were to be found throughout the region. Among those of note were William B. Rogers in Virginia, Elisha Mitchell and Moses A. Curtis in North Carolina, Gerard Troost in Tennessee, Charles W. Short in Kentucky, Alvan W. Chapman in Florida, John L. Riddell in Louisiana, and Gideon Lincecum on the Texas frontier.

The greatest activity in natural history was concentrated in the Charleston area. Long the chief center of southern science, this city and its environs were home to some of the most outstanding scientific figures of the Old South, such as Stephen Elliott, John Bachman, John E. Holbrook, and Henry William Ravenel. So many active students of science made Charleston the Old South's most important scientific community. This remarkable group's pursuit of science was nurtured by the Charleston Museum, one of the nation's oldest and most important collections of natural history specimens, and the Elliott Society of Natural History, one of two noteworthy scientific societies in the Old South, the other being the New Orleans Academy of Sciences.

When compared with its outstanding performance in natural history, the Old South's showing in the pure sciences was strikingly lackluster. This situation was the result of a combination of factors: the absence of a tradition of important activity and accomplishment in pure science, the low level of professionalization and institutional development that characterized southern science, and cultural considerations. The cumulative effect was the perpetuation of the South's preoccupation with the collection, description, and classification of natural phenomena and the relegation of experimental research to the periphery of scholarship.

Like Americans everywhere, antebellum southerners were keenly interested in the practical applications of science. Applied science in the Old South largely involved attempts to bring science to bear on the region's mounting agricultural problems toward the end of the era. The highly acclaimed research of Edmund Ruffin in soil chemistry is a case in point.

Like science, medicine in the antebellum South exhibited unmistakable regional characteristics. By the emergence of the Old South, a distinctive southern health picture was evident. It was the worst in the nation. So poor was the state of health in the region that northern life insurance companies charged their southern policyholders higher premiums. Malaria remained endemic and was the principal cause of disability and death. Residents of the southern port cities and the surrounding countryside lived in fear of yellow fever, which became a southern disease in the 19th century. New Orleans, the Old South's largest city, was popularly known as "the graveyard of the Southwest" because of its frightful mortality rate (nearly three times that of Philadelphia and New York). Infant mortality rates in the South were the highest in the nation. In addition, it is estimated that as many as half of all southern children suffered from hookworm infection, a condition not diagnosed until the opening years of the 20th century. Finally, inadequate diets, poor housing, unhealthy quarters, and hazardous working conditions exacted a heavy toll on the health of the South's large slave population.

The Old South's health problems were the result of environmental and cultural factors. Climate and frontier conditions in the developing region, in conjunction with slavery, combined to account for the continued presence of malaria. The insect vectors of yellow fever and typhoid fever also thrived. In addition to fostering insect life, the long, hot summers made the preservation of food difficult, increased sanitary problems, and encouraged going barefoot, a habit associated with the spread of hookworm.

The growing cultural lag that increasingly set the South apart from the more progressive North contributed to regional health problems in a variety of ways. The low level of southern education, the lowest in the nation, clearly complicated the health picture. Nationwide, the "heroic" procedures of physicians were questioned during the first half of the 19th century, a development that encouraged the reliance on traditional healers and self-dosage with patent medicines. The rural and undereducated southerners were particularly prone to resort to these health-threatening practices. The absence of a social conscience on the part of the dominant planter class also had an adverse effect on health. Finally, institutions for the advancement of medicine, such as schools and journals, were, like those in science, slow to appear in the overwhelmingly rural South, and those that were founded faced a difficult struggle for survival. The few that did survive were inferior to those in the North. The rise of southern medical nationalism, or states' rights medicine, did little, despite its rhetoric, to change this situation. The product of regional patterns of disease and sectional tensions, states' rights medicine stressed the uniqueness of the South's medical problems and the subsequent need for southern-trained physicians and a southern medical literature. Although the desire to improve the practice of medicine in the region was indisputably one of its goals, southern medical nationalism, like the scientific racism of Josiah C. Nott and others, was primarily a defense of the civilization of the Old South. Consequently, it contributed more to sectionalism than to medical reform.

Although easily overlooked because of the health problems of the region, the antebellum South's strides in surgery contributed significantly to the rise of modern medicine. Two southerners—Ephraim McDowell and J. Marion Sims—achieved international acclaim in operative obstetrics and gynecology. The former, while practicing on the Kentucky frontier in 1808, performed the first successful ovariotomy, pioneering abdominal surgery. The latter, an Alabama surgeon, used slave women as subjects to perfect, in the 1840s, the initial procedure for the treatment of vesicovaginal fistula, a major breakthrough in gynecology. The third southerner who contributed to the birth of American surgery was Crawford W. Long, a small-town Georgia physician who was the first to use ether as a surgical anesthesia in 1842, helping to launch a new age of painless surgery.

Civil War. The culture of the Old South was inhospitable to scientific inquiry and threatening to health, but the region's scientists and physicians closed ranks with their countrymen to defend it against all perceived enemies. In 1861, when the South withdrew from the Union, they pledged their lives and fortunes to the new Confederate nation.

The Civil War was not a scientific war. Neither the North nor the South used scientific talent in ways that led to new or drastically improved weapons that altered tactics and strategy on the battlefield. Still, each side made extensive use of scientists. The North did considerably better than the South in this area. Prominent scientists were consciously incorporated into the northern war effort in an advisory capacity. In the South, they were engaged as problem solvers. The Confederacy's failure to devise a science policy is attributable to the many and pressing problems to be overcome in order to wage war and the popular perception of scientists as problem solvers in the South.

Scientists in the southern war effort worked in the government-run munitions industry, primarily in the War Department's Ordnance Bureau. Headed by Josiah Gorgas, this agency was responsible for the Confederacy's supply of war materiel. Gorgas and his assistants accomplished the near impossible, building a munitions industry from scratch. It was largely through their efforts that the South was able to keep its armies in the field for four years against a vastly superior enemy. Indeed, the Confederacy ran out of men before it did arms.

Disease, disability, and death stalked the Civil War soldier and made this the costliest conflict in American history. A major reason for the unequaled carnage was the state of contemporary medicine. Indeed, the nation's doctors were plunged without warning into a modern war with its unprecedented medical problems at a critical turning point in American medical history. Out of this era of transition, which saw established beliefs and practices come under attack, was to emerge the beginning of modern American laboratory medicine. In the meantime, a majority of the standard therapeutic measures—puking, purging, bleeding, and giving large doses of potentially dangerous drugs in particular—met with little success in the day-to-day struggle against common complaints and failed miserably when confronted by yellow fever, cholera, and typhoid fever, the great killer epidemics of 19th-century America.

The cruelest blow of all to the Civil War soldier was that the life-saving antiseptic management of wounds, growing out of the research of Pasteur and Lister, came too late to be of help. Consequently, any serious injury to a limb meant amputation and the distinct possibility of death from one of the so-called surgical fevers—gangrene, erysipelas, or pyemia. Abdominal wounds were especially feared and constituted an almost certain death sentence.

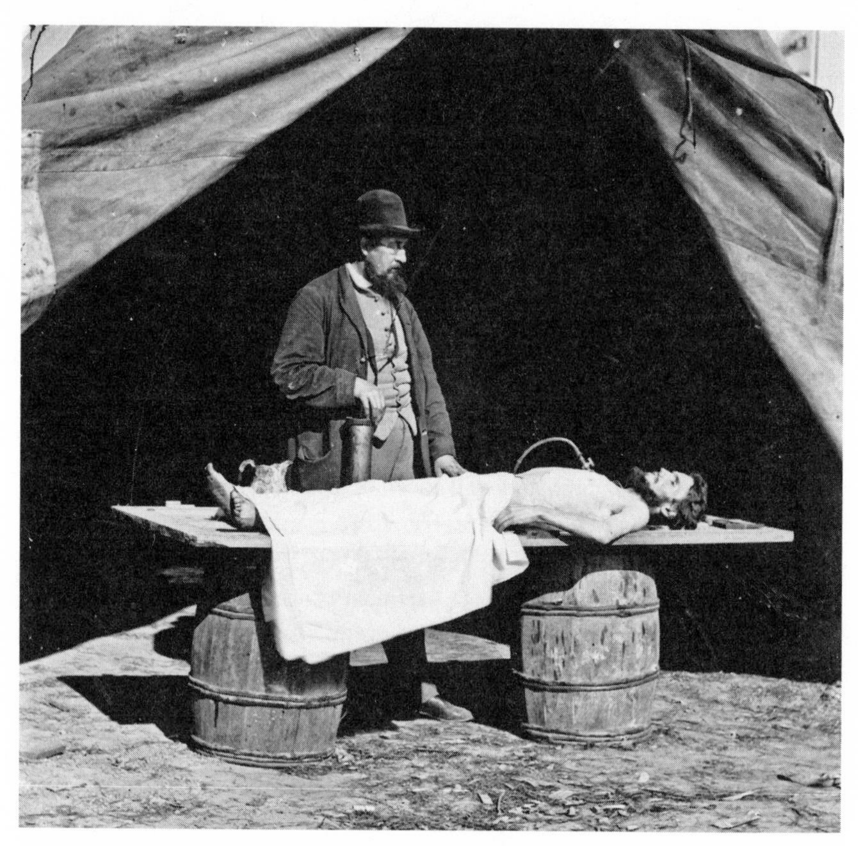

Embalming surgeon at work on soldier's body (Photographer unknown, Library of Congress [LC-B811-2531], Washington, D.C.)

The Civil War's toll of misery and death was most evident in the Confederate army. Like Gorgas and the manufacture of the tools of war, Samuel P. Moore, the southern surgeon general, had to build a medical service from scratch. He also was successful in meeting this challenge. But the efforts of the Confederate medical officers were hampered by an inadequate supply of trained physicians, near-crippling shortages of medicines and medical stores, and a steadily worsening military situation. Still, they faithfully kept at their tasks and provided valiant and commendable service. Few medical lessons, however, emerged from the carnage. For the South, the chief gain was the experience that the war provided in the treatment of the sick and injured and most especially the sharpening of surgical skills.

New South. The civilization of the Old South perished on the battlefields of the Civil War. The legacy of the South's failed bid for independence was frustration, poverty, and obsession with the past. These things supplanted such long-standing cultural determinants as agrarianism and ruralism in significance. Their immediate and lasting effect was to blight life in the region. Indeed, passing time seemed only to worsen matters. Even the much ballyhooed New South movement, with its promise of progress and prosperity based on the northern industrial model, did little to relieve the plight of the southern people. Consequently, as the New South era drew to a close in the opening years of the 20th century, the South was mired in backwardness and misery. Nowhere is this better seen than in science and medicine.

In the North, the Civil War was a catalyst for scientific progress, and American science matured rapidly during the last years of the 19th century, paving the way for domination in the century ahead. In the South, the war produced intellectual stagnation. Consequently, southern scientific leaders were largely sideline observers of Gilded Age and early 20th-century advances in science. The comparative gap separating southern and national science was probably greater at the end of the New South period than ever before.

With life in the South reduced to a scramble for survival and with spiritual malaise rampant in the dark days after Appomattox, there was little opportunity or desire to pursue science. So unbearable were conditions to some scientists, including John and Joseph LeConte, two of the New South's emerging scientific leaders, that they joined the postwar exodus from the region. Yet the beginnings of the revival of southern science date from the immediate postwar period. As the initial shock and agony of defeat began to abate, southern scientists reestablished contact with northern friends, who generously assisted them in rekindling their scientific interests, providing, for example, news of wartime developments in science and copies of recent works.

The first tangible signs of the revival of southern science were predictably found in natural history. The principal figures, men like Henry William Ravenel, Moses A. Curtis, and Alvan W. Chapman, were holdovers from the antebellum period. Natural history would continue to dominate science in the South, but its heyday was at an end. It was giving way to the modern science of botany, and the evolving professional scientist was supplanting the amateur collector.

As before, the southern record in the pure sciences was meager and undistinguished. The fate of this branch of science was inextricably bound up with that of higher education. At the beginning of the period, the South's colleges and universities were paralyzed by the effects of the Civil War and the political turmoil of Reconstruction. By its end, the best of them had made the transition to multipurpose institutions. But the region's continuing poverty, conservative social philosophy, and religious fundamentalism prevented them from becoming true centers for the advancement of learning.

The prospects of applied science in the South were as encouraging as those of abstract science were unpromising. Economic necessity made this the case. A long list of southern scientists sought to bring science to bear in the effort to restore the region's prosperity. In agriculture, the traditional but troubled source of southern wealth, George Washington Carver exemplifies the renewed emphasis on scientific farming and farm management. In industry, Charles Holmes Herty's research on forest products is illustrative of the attempt to use science to create new economic opportunities.

The Civil War and its aftermath had a disastrous effect on health in the South. On the one hand, the hostilities left untold thousands of southerners in precarious or weakened health. On the other, the conflict's legacy of poverty exacerbated the region's tradition of poor health. As a result, old diseases increased in incidence and virulence, and new health problems arose. Malaria, the leading cause of debility and loss of efficiency in the antebellum South, had showed signs of decline in the decade preceding the Civil War. In the postwar period, however, it soared to record levels and reappeared in areas where it had previously been brought under control. Yellow fever, another old and distinctively southern disease, was a recurrent source of terror, death, and economic blight. Tuberculosis was more prevalent in the South than elsewhere. Blacks were especially hard hit. But a higher incidence of tuberculosis was only one indication of the deteriorating health of the former slaves. Left to fend for themselves after the collapse of Reconstruction, freedmen experienced excessively high rates of sickness and death.

Black health problems, as well as those of a growing number of whites, were in large part the result of the rise and spread of tenancy, the cruel backbone of postwar southern agriculture. The proliferation of the mill town, the chief symbol of the New South, further eroded the health of the poor whites. The principal diseases of poverty were hookworm and pellagra. The former, although undetected, was an old health hazard. In the antebellum period, however, it had been limited to slaves and the relatively small class of poor whites. Postwar poverty exposed growing numbers of southerners to hookworm infection, making it a major threat to regional health. Pellagra was the most spectacular and deadly of a number of disorders caused by dietary deficiency that plagued the swelling ranks of the southern poor. Almost exclusively southern in incidence, hookworm and pellagra were widespread by the time they were diagnosed at the turn of the century.

Urbanization in the South was largely a postbellum phenomenon, and New

South cities were notoriously unhealthy. Chief among the health hazards were unpaved and poorly drained streets, inadequate or nonexistent sewage arrangements, public garbage heaps, and contaminated water supplies. Conditions were the worst in the segregated quarters into which urban blacks were crowded. With their growth came the multiplication of health problems. Health administrations were virtually nonexistent before the last two decades of the century, so little was done to improve conditions.

The threat of disease did not go unnoticed in the South. But poverty and the inability of the medical profession to combat the principal causes of morbidity and mortality stymied would-be reformers. Toward the end of the century, however, improvement in the southern economy and the acceptance of the new germ theory of disease provided the opportunity for health reform. The result was the genesis of the southern public health movement. Boards of health were established and empowered to investigate and combat health problems. Although the effectiveness of these agencies was limited by inadequate budgets, legislative interference, suspicion and hostility on the part of businessmen, and the ignorance of the masses, they pushed health reform on a broad front. The state boards of health uncovered and attacked a host of health hazards, inspected water supplies, sought to impose quarantines during outbreaks of epidemic disease, supervised vaccines, published reports, and strove to educate the public on health matters.

20th-Century South. Still suffering from the physical and emotional effects of the Civil War, the South entered the 20th century with an uncertain future. The years ahead, however, brought slowly improving prospects, the result of the slow transformation of southern society during the interwar years. Indeed, change became the major theme of southern history. The Progressive movement, which flowered after 1910, marked a turning point in the region's reaction to change by making it palatable, indeed desirable, to many southerners. Following World War I, change resumed, stronger than ever.

To the region's traditionalists, however, the prospects of change not only were subversive to the southern way of life but actually threatened to destroy it. The result was social controversy that set southerner against southerner. The struggle for control of the South's destiny was long and often bitter, but the outcome was inevitable—ever so slowly there was an erosion of the South as a distinctive social and cultural entity. Put another way, the South was closing the circle, gradually moving back into the national mainstream, which it had left during the era of the Old South. The consequences of the Americanization of the South were to be immensely beneficial for science and medicine, although they were not to be realized until after World War II. In the meantime, each faced continued rough sledding.

The maturation of American science accelerated during the interwar years, and by the outbreak of World War II, the United States was poised for world scientific leadership. Owing to its physical, cultural, and intellectual poverty, the South contributed only marginally to national greatness in science.

The state of science during this transitional period in the South's history is best seen through an examination of academic science, for scientific inquiry nationwide was largely university centered. Building on the beginnings from the Progressive Era, southern education, from bottom to top, underwent progressive change after 1920. By the outbreak of World War II, the South had experienced a veritable revolution in education. The prospects for the region's colleges and universities, however, were not as bright as the foregoing might seem to indicate. The gap to be closed was great and progress was slow and uneven. Indeed, the perennial problems facing southern schools were numerous and weighty. They included inadequate financial resources, overworked and underpaid faculties, mediocre students, and the desire of communities to control their schools. Such factors severely limited the capacity of the southern schools for intellectual attainment.

The South's institutions of higher learning, however, were not devoid of scholarship. As a matter of fact, there was a general reawakening of the southern intellect in the 1920s. But science fared less well than the social sciences and humanities in the southern intellectual renaissance. To be sure, the South had dedicated scientists. They worked to keep up with developments in their fields of interest, and some engaged in research, the results of which were occasionally noteworthy. Moreover, with the gradual upgrading of student bodies, faculties, and facilities as the 20th century wore on, the general state of science in the region improved. Still, when viewed comparatively, science in the South's institutions of higher learning was undistinguished.

As before, the realities of southern life shed informative light on the reasons for the state of science in the region. The interwar years were the era of the Benighted South. This popular image was the product of the South's longstanding social and economic problems and the unprecedented attention that the raging controversy between the proponents of change and the traditionalists drew to them. Benightedness influenced science in two major ways. First, the region's continuing economic problems, which worsened in the 1920s, meant modest expenditures on education, thereby limiting what could be done in science. Science also found itself embroiled in a conflict with the forces of social and religious fundamentalism. Such a confrontation was perhaps inevitable, for the vital interests of the two were diametrically opposed—free inquiry on the one hand and intellectual conformity on the other. The battleground was evolution, and the celebrated Scopes Trial of 1925 was only the best known of a host of violations of freedom of thought by the crusading fundamentalists. But the proponents of progress refused to yield to the traditionalist onslaught and tenaciously resisted. Their most powerful weapon, however, was time; for try as they might, the traditionalists could not quarantine the South from the formidable and unrelenting winds of change that were buffeting the region.

Owing to a series of medical discoveries around the turn of the century that propelled the southern public health movement into a new stage of activity and accomplishment, medicine made greater progress than science in the South after 1900. Between 1898 and 1906, the insect carriers of malaria (1898) and yellow fever (1899) were identified, and hookworm (1902) and pellagra (1906) were diagnosed as endemic among the southern poor. On the one hand, these developments vividly underscored the South's unique and stigmatizing health problems and focused national attention on them. On the other, they paved the way for the eventual control of the region's principal causes of sickness and death and promoted increased interest in public health reform.

The campaigns against malaria, yellow fever, hookworm, and pellagra, although hindered by regional poverty and the resistance of business and political leaders who were outraged over the embarrassing exposure of the South's myriad and frightful health problems, were landmark victories for southern health. By the end of World War II, these scourges had been eradicated (or controlled in the case of pellagra), in large part as the result of the national discovery of the South's health plight after 1900. While publicity about the shocking state of southern health reinforced the stigma of regional backwardness, it also led to crucial assistance from northern philanthropies and the federal government. The indispensable roles of the Rockefeller Foundation in the control of hookworm and the U.S. Public Health Service in the fight against pellagra are cases in point.

The nascent southern public health movement was a major beneficiary of the late 19th- and early 20th-century medical advances that stripped the region's principal diseases of their mystery. As increasing numbers of southerners became aware of the modern concept of disease and the life-saving potential of laboratory medicine, the long-standing belief that an unhealthy climate was the cause of disease was toppled, the importance of sanitation and drainage was recognized, and a lessening of opposition to the recognition of regional health problems and a greater willingness to confront them evolved. These developments coincided with and were influenced by the southern Progressive movement. Chagrined by the South's backward image, the Progressives sought to rid the region of the principal causes of backwardness. Health reform was high on their agenda.

Southerners attacked disease on a broadening front. Crusades against malaria, yellow fever, hookworm, and pellagra touched off similar campaigns against tuberculosis and syphilis. Communities built sanitaria and hospitals. Those states that had not already established them set up boards of health. Responding to the stimulus of the Rockefeller Sanitary Commission for the Eradication of Hookworm Disease, which combated this disease at the local level, county health departments mushroomed, propelling the South into the lead in this area. And health department expenditures increased, despite the region's ongoing economic woes. Additional funding for health reform came from philanthropic organizations and federal agencies. The Frontier Nursing Service, established in 1925 by Mary Breckinridge in the mountains of Kentucky, typified the growing concern for the health of the isolated people of southern Appalachia. Finally, southern senators and congressmen began to take a greater interest in health legislation. The cumulative effect of these developments was the gradual improvement of southern health. The narrowing of differences in mortality rates between the regions attested to the gains made.

But as revolutionary as the progress in health reform was, the South remained the nation's sickliest section at the onset of the Depression. Familiar disease forms continued to plague the region. For example, malaria had not been brought under control, and the plummeting of cotton prices in the 1920s led to a resurgence of pellagra. And southern cities remained unhealthy.

The first attempts at a national health program were made during the New Deal, and the South was a major beneficiary of the New Dealers' concern for health. Funds for medical care were provided by the Federal Emergency Relief Administration. Civilian Conservation Corps members received medical attention. The draining of 2 million acres of swamp by the Civil Works Administration, the Federal Emergency Relief Administration, and the Works Progress Administration and studies of the breeding habits of mosquitoes by the Tennessee Valley Authority expedited efforts to eradicate malaria. The control of typhoid fever and dysentery was advanced through the federally sponsored construction of 2.3 million sanitary privies by 1939. New crusades against tuberculosis and venereal disease were launched. The Works Progress Administration built hospitals and sewage plants. The Federal Housing Administration's slum-clearance programs, half of which were in the South, promoted urban health reform. Of far-reaching significance was the Social Security Act

Physician visiting a patient and her family in a needlework scene by Ethel Mohamed, photographed in 1978 (Jane Mosley, photographer, Center for Southern Folklore, Memphis, Tenn.)

of 1935. This historic piece of social welfare legislation provided federal funds for health purposes and created permanent machinery for distributing them.

World War II had major uplifting effects on southern health. The stationing of large numbers of troops in the region brought the resources of the federal government to bear in the fight against disease on an unprecedented scale. As the conquest of malaria illustrates, public health was greatly advanced. The health screenings and medical attention that accompanied military service led to vastly improved health for thousands of southerners. And military instruction in hygiene inculcated in them the importance of good health and taught them how to achieve it. The wartime appearance of enriched flour and bread, containing synthetic vitamins, considerably curtailed the threat of pellagra and other disorders resulting from dietary deficiencies.

World War II was a landmark in southern history: it reinvigorated the region's long-troubled economy and swung the battle between the modernists and traditionalists in favor of progressive change, greatly accelerating the Americanization of the South. Indeed, no other period in the South's past witnessed as much fundamental change as the years after World War II. The changes in southern life extended across a broad front. Prominent among them were the triumph of industry, the transformation of agriculture, burgeoning urbanization, the breaking of the hold of ruralism, the ending of physical and cultural isolation, the dismantling of the Jim Crow system, the disintegration of the political Solid South, and a revitalized role in national politics. The result, as one historian put it, was the demise of the "sectional South" and the rebirth of the "American South." Although accomplished at the expense of regional distinctiveness and at times, as in the case of race relations, vigorously opposed, the Americanization of the South led to unprecedented prosperity, dramatic improvements in the quality of life, growing opportunities for southerners, and renewed respect for the region. Science and medicine were major beneficiaries of the postwar changes in southern life.

Recent South: Science. The Americanization of the South in the mid-20th century sparked a veritable revolution in the pursuit of science in the region. It benefited from the increased prosperity, public policy focus on economic development, growth of the middle class, and maturation of southern universities. To be sure, popular attitudes can be resistant to scientific consensus, as in the continued opposition of many southerners to the teaching of evolutionary ideas in schools. A Southern Focus Poll in 1998 showed that 67.7 percent of southerners believed the Genesis account of creation in the Bible, compared to 17.1 percent who accepted Darwin's theory of evolution. Nonsoutherners were twice as likely to believe in Darwin's ideas compared to southerners. Fourteen states in the 1990s faced controversies over evolution and creationism—the claim of scientific proof for the biblical account of the origin of life. The teaching of creationism brought numerous court cases and divisive controversies, but, more recently, ideas of "intelligent design" have served as an alternative to Darwinian evolution by not mentioning God or religious belief while maintaining that the complexity of the universe suggests a transcendent force behind creation.

The authority of science in general has nonetheless dramatically increased in the past half century. At the outbreak of World War II, the South was little more than an outpost of American science. In the aftermath of the war, it rose to the status of junior partner in the national scientific establishment. This amazing turnabout is easily demonstrated. The South produced and used scientists at greater rates than ever before in its history. Institutions of every type for the support of science became unprecedented in quantity and quality. Levels of scientific achievement soared. The region's attitude toward science became increasingly progressive and supportive. The great distance that science in the South traveled after World War II was clearly seen in the reversal of the South's "brain drain"; the region became a magnet for scientific talent.

Much of the authority of science has come from its relationship to economic development and technological advancement through research. Justifications of government funding for research have been framed in terms of economic benefit, often tied to technological education. Honda, Toyota, Nissan, Mercedes-Benz, BMW, Volkswagen, and Isuzu have all established automobile factories in the South in the past two decades, and the quality of the postsecondary educational infrastructure in site locations has been used as a recruitment tool, often requiring Deep South states to put more funding into not only universities but K–12 schools. States have strengthened community and junior colleges through "Workforce Education" and "Career and Technological Education" divisions that teach technical skills. The University of Alabama and Mississippi State University have established institutes dedicated to research programs aimed at automobile manufacturing innovation. The state of Arkansas's Science and Technology Authority encourages research and development projects, business innovation, and education in science, mathematics, and engineering. The Tennessee Technology Corridor in the eastern part of the state coordinates efforts to recruit high-tech companies into the area. Recent discussions of educational improvement at the K–12 level have included efforts to strengthen school curricula in science and technology.

The stress on scientific and technological research has been important in giving the South the nation's highest rate of economic expansion in the past three decades, seen not only in automobile factories but in growth in agribusiness, defense industries, and energy resources. Changes in southern society since the 1960s led to new opportunities for utilizing the talents of women, African American, and Latino scientists in the region. Southern universities have embraced new missions of contributing research on a wide range of interests but geared toward economic development in particular. Eleven southern universities are among the nation's 59 universities that are members of the respected Association of American Universities (AAU). Sixteen other southern universities are recognized as major research institutions because of their ability to attract research grants and confer significant numbers of doctoral degrees. This growth in research universities is striking because it is relatively recent—at the end of World War II only five southern universities were members of the AAU. Further illustrating the South's competitiveness in research is the selection of six southern institutions to participate in President Barack Obama's Advanced Manufacturing Partnership, bringing together industry, the federal government, and leading research institutions to promote such emerging technologies as robotics, information technology, biotechnology, and nanotechnology.

Collaborative research institutes have been important parts of the growth of science and technology in the recent South. North Carolina's Research Triangle Park is the world's largest industrial research complex, with 7,000 acres that contain industrial laboratories, trade associations, federal and state government research facilities, nonprofit research centers, and university-related research activities. Duke University, the University of North Carolina at Chapel Hill, and North Carolina State University provide the "triangle" of interrelated, collaborative engagement. More than 700 research organizations are represented, with 42,000 full-time employees, working in such research areas as environmental sciences, life sciences, pharmaceuticals, agricultural chemistry, biotechnology, microelectronics, and information technology. The Southern Research Institute is another longtime research center in the South, established in Birmingham, Ala., in 1944, and its scientists have contributed to advances in, among other areas, cancer research, the Space Shuttle program, and the chemical industry. Alabama is also home to the Cummins Research Park, the second-largest research and technology park in the United States and the fourth-largest such complex in the world. Hosted by the University of Alabama in Huntsville, Cummins has 250 companies on site, ranging from telecommunication companies to biotechnology laboratories to aerospace and defense firms. Dating from the development of the post–World War II Redstone Arsenal with its German rocket engineers, Cummins was established in 1962 and now covers 3,840 acres with more than 11 million square feet of space for research, development, and manufacturing. Georgia established a similar institute, the Georgia Research Alliance, in 1990 as a nonprofit corporation headquartered in Atlanta. Its Eminent Scholars initiative recruits distinguished scientists from around the world to direct research and development projects with economic benefit to the state.

The information technology revolution has given individuals inexpensive ways toward new kinds of research through personal innovation, creativity, and collaboration through technology, and people in the South take part in such activities. Hand-held computers, social networks, and what has been called "the cloud," which stores applications that anyone can download, play a part in making individuals capable of imaginative leaps. Globalization results in integrating these empowered individuals into ecosystems where they create new products and provide services in creative ways. Such empowerment knows no regional lines. These ecosystems may, however, come together in certain particular places that combine universities, an educated populace, dynamic business sectors, and extensive broadband connections. One southern place, Austin, Tex., has been identified as such a site, along with Silicon Valley, Calif., Cambridge, Mass., Boulder, Colo., and Ann Arbor, Mich.

Recent South: Medicine. The postwar transformation of the South also brought unparalleled improvements in health. Indeed, the persistent image of a unique "sickly South" became a thing of the past as the region moved toward national patterns and norms in health matters. In medicine, the region's infrastructure advanced markedly, with health-related institutions taking their place as competitive with any in other places. Health problems that have regionally distinctive dimensions remain, though, many of them legacies of the South's historic poverty, racial divisions, and rural isolation. As before, ethnic minorities and the poor are the least healthy of southerners.

Federal, state, and private funding promoted a dramatic expansion of health and biomedical institutions over the past several decades. Advances in drugs and medical technology and the increased number of professional health personnel have led to the expansion of hospitals and clinics and increased lifespans, but also increased costs for medical care. HMOs have become a key to the health delivery system. The economic importance of the health-care industry to the South is seen in the higher proportion of the region's work force involved in health care than is the case in other parts of the United States. Urban areas are usually the locus of the largest medical centers, which often serve as major icons in southern cities: St. Jude Children's Research in Memphis, the Centers for Disease Control and Prevention in Atlanta, the M. D. Anderson Cancer Center in Houston, and the University of Alabama at Birmingham Medical Center. In earlier decades, urban medical centers served a large radius of small towns and rural areas that often did not have hospitals. Today, secondary medical institutions in small cities and towns have excellent facilities.

As in the rest of the nation, the South's poor and underemployed working people cannot afford health insurance, and reforms in the nation's health-care system, promoted by President Obama and passed by Congress in 2010, require Americans to have insurance as part of new state-administered health systems. Public health agencies, community-based organizations, and grassroots movements all provide services and advocacy for those needing health care. Physicians are a linchpin of the South's health-care system, working in individual and group private practices, rural clinics, hospitals, nursing homes, health departments, and health corporations. The small-town country doctor was once a defining figure in the region's medicine, but the diversity of care options is the new norms. Most southerners become familiar with hospitals, where specialty care can be provided, emergencies cared for, and operations done. A physician shortage characterizes many places in the South, especially in rural areas.

One medical procedure has been the center of a political firestorm, with the Supreme Court's *Roe v. Wade* (1973) decision prompting conservative religious activists to argue for reform in the public's access to the procedure since then. Abortion numbers have declined since 1991, with the southern states mostly sharing in that national trend. Southern states have some of the strongest restrictions in the nation on abortion. Few clinics exist in the South to perform the procedure. Access to legal abortion has steadily declined as state legislatures have reduced public funding for abortion.

The Centers for Disease Control and Prevention in Atlanta reports that the southern states in the U.S. Census Region have higher disease rates than the rest of the nation for such chronic diseases as lung cancer, diabetes, stroke, heart disease, and obesity. African Americans are especially at risk in the region, with disproportionately high rates of asthma, diabetes, obesity, infant mortality, and HIV/AIDS infections. The Deep South states rank poorly on most health indicators, but even a border state like Tennessee is notable nationally for poor health statistics. Florida and Virginia, on the other hand, show health indicators more in line with national trends. Authorities attribute the poor health picture in the region overall to poverty, lack of access to primary care, poor nutrition, and sometimes toxic environmental conditions.

The South has the nation's highest rates of rural poverty, where many health problems persist. High motor vehicle death rates and child death rates probably reflect rural conditions, where inadequate public transportation, limited access to health-care facilities and decent housing, and lack of low-cost, healthy food contribute to specifically rural high disease rates. The region's urban areas are sometimes not much better. One index of health status ranks only three southern cities as among the nation's 25 healthiest places, with 13 southern urban areas ranking among the nation's least healthy. Inner cities are particularly unhealthy places, correlating with concentrations of low-income minority populations. In the last decade, the arrival of immigrants, especially Latinos, in significant numbers in the Deep South and the Atlantic South has put new strains on the South's health-care systems, creating the need for health-care providers to respond to different languages, cultural expectations, and lifestyles.

Several health conditions continue to characterize the South as a whole. Infant mortality is one of the most persistent health issues the region has faced. Infants born in the South are more likely to be born prematurely, to be low weight, or die as infants. The infant mortality rate declined in the first decade of the 21st century but has risen most recently. Obesity is another health problem that characterizes the region, with almost 80 percent of counties in the South having high rates of obesity and related diabetes. One study found that 9 of 10 states with the highest rates of adult and childhood obesity are in the South, and the region leads the nation with the highest percentages of high blood pressure and diabetes. Poor people in rural areas are particularly at risk for obesity, often living in "food deserts," defined as areas with limited access to conventional grocery stores. In one extreme example, Quitman County in Mississippi has only one grocery store in the county. Finally, HIV/AIDS dates in the nation to the 1960s but was not diagnosed as a disease crippling the immune system until the early 1980s. By 2010 the disease had killed 600,000 Americans, including 230,000 in the South. AIDS organizations first appeared in the South in 1982, and southern researchers made important contributions to the diagnosis and treatment of the disease. By the early 2000s the disease had taken deep root in the South, with 7 of the 10 states with the highest AIDS case rates being in the region.

In retrospect, southern science and medicine have closed the circle. They began on a roughly equal footing with the rest of British North America. With the rise of a distinctive southern culture after the Revolution, they became sectional and second-rate. The Civil War and its lingering aftermath perpetuated the South's scientific lag and poor health well into the 20th century. But since World War II national patterns and norms have increasingly prevailed. The

South has made enormous strides as a place valuing research, and its health-care system has a strong infrastructure, although inequities remain, and particular diseases continue to haunt the region's people.

JAMES O. BREEDEN
Southern Methodist University
JAMES G. THOMAS JR.
CHARLES REAGAN WILSON
University of Mississippi

Amanda Carson Banks, *Birth Chairs, Midwives, and Medicine* (1999); Sandra Lee Barney, *Authorized to Heal: Gender, Class, and the Transformation of Medicine in Appalachia, 1880–1930* (2000); Wyndham B. Blanton, *Medicine in Virginia in the Eighteenth Century*, 3 vols. (1930–33); Erica Brady, *Healing Logics: Culture and Medicine in Modern Health Belief Systems* (2001); James O. Breeden, *Joseph Jones, M.D.: Scientist of the Old South*; Robert D. Bullard, *Dumping in Dixie: Race, Class, and Environmental Quality* (1990); Clark R. Cahow, *People, Patients, and Politics: A History of North Carolina Mental Hospitals, 1848–1960* (1982); Gregory Cajete, *Native Science: Natural Laws of Interdependence* (1999); James H. Cassedy, *Journal of History of Medicine and Allied Sciences* (April 1973), *Medicine in America: A Short History* (1991); Anthony Cavendar, *Folk Medicine in Southern Appalachia* (2003); James X. Corgan, ed., *The Geological Sciences in the Antebellum South* (1982); Horace H. Cunningham, *Doctors in Gray: The Confederate Medical Service* (1958); Pete Daniel, *Toxic Drift: Pesticides and Health in the Post–World War II South* (2005); George Daniels, *American Science in the Age of Jackson* (1968); David Dary, *Frontier Medicine: From the Atlantic to the Pacific, 1492–1941* (2008); Richard Beale Davis, *Intellectual Life in the Colonial South, 1585–1763*, 3 vols. (1978); William H. Deaderick and Lloyd Thompson, *Endemic Diseases of the Southern States* (1916); David H. DeJong, *If You Knew the Conditions: A Chronicle of the Indian Medical Service and American Indian Health Care, 1908–1955* (2008); John Dittmer, *The Good Doctors: The Medical Committee for Human Rights and the Struggle for Social Justice in Health Care* (2009); Gregory Dorr, *Segregation's Science: Eugenics and Society in Virginia* (2008); John Duffy, *Epidemics in Colonial America* (1953), *The Healers: A History of American Medicine* (1976), *Journal of Southern History* (May 1968), ed., *The Rudolph Matas History of Medicine in Louisiana*, 2 vols. (1958–62); Clement Eaton, *The Mind of the Old South* (1964); Clark A. Elliott, *Biographical Dictionary of American Science: The Seventeenth through the Nineteenth Centuries* (1979); Elizabeth W. Etheridge, *The Butterfly Caste: A Social History of Pellagra in the South* (1972); John Ettling, *The Germ of Laziness: Rockefeller Philanthropy and Public Health in the New South* (1981); Sharla M. Fett, *Working Cures: Healing, Health, and Power on Southern Slave Plantations* (2002); Michael A. Flannery, *Civil War Pharmacy: A History of Drugs, Drug Supply and Provision, and Therapeutics for the Union and the Confederacy* (2004); Gaines M. Foster, *Journal of Southern History* (August 1982); Brooke Hindle, *The Pursuit of Science in Revolutionary America, 1735–1789* (1956); Howard L. Holley, *A History of Medicine in Alabama* (1982); Pippa Holloway, *Sexuality, Politics, and Social Control in Virginia, 1920–1945* (2006); Margaret Humphreys, *Malaria: Poverty, Race, and Public Health in the United States* (2001), *Yellow Fever and the South* (1992); Stephen Inrig, *North Carolina and the Problem of AIDS: Advocacy, Politics, and Race in the South* (2001); James H. Jones, *Bad Blood: The Tuskegee Syphilis Experiment* (1993); Alan M. Kraut, *Goldberger's War: The Life and Work of a Public Health Crusader* (2003); Edward Larson, *Sex, Race, and Science: Eugenics in the Deep South* (1995), *Summer for the Gods: The Scopes Trial and America's Continuing Debate over Science and Religion* (1997); Dorothy Long, ed., *Medicine in North Carolina: Essays in the History of Medical Science and Medical Service, 1524–1960*, 2 vols. (1972); Deborah Kuhn McGregor, *Sexual Surgery and the Origins of Gynecology: J. Marion Sims, His Hospital, and His Patients* (1990); Sally G. McMillen, *Motherhood in the Old South: Pregnancy, Childbirth, and Infant Rearing* (1990); Edward T. Martin, *Thomas Jefferson, Scientist* (1952); Nancy Smith Midgette, "The Role of the State Academies of Science in the Emergence of the Scientific Profession in the South, 1883–1983" (Ph.D. dissertation, University of Georgia, 1984); Steven Noll, *Feeble-Minded in Our Midst: Institutions for the Mentally Retarded in the South, 1900–1940* (1995); Ronald L. Numbers, *The Creationists: The Evolution of Scientific Creation* (1992); Ronald L. Numbers and Janet Numbers, *Journal of Southern History* (February 1982); Ronald L. Numbers and Todd L. Savitt, *Science and Medicine in the Old South* (1989); Nathan Reingold and Marc Rothenberg, eds., *Scientific Colonialism, 1800–1930: A Cross-Cultural Comparison* (1986); Todd L. Savitt, *Journal of Southern History* (August 1982), *Medicine and Slavery: Diseases and Health Care of Blacks in Antebellum Virginia* (1978), *Race and Medicine in Nineteenth- and Early Twentieth-Century America* (2007); Glenna R. Schroeder-Lein, *Confederate Hospitals on the Move: Samuel H. Stout and the Army of Tennessee* (1994); Marie Jenkins Schwartz, *Birthing a Slave: Motherhood and Medicine in the Antebellum South* (2006); Richard H. Shyrock, *South Atlantic Quarterly* (April 1930); Susan L. Smith, *Sick and Tired of Being Sick and Tired: Black Women's Health Activism in America, 1890–1995* (1995); Loudell Snow, *Walkin' over Medicine: Traditional Health Practices in African-American Life* (1993); Raymond Phineas Stearns, *Science in the British Colonies of North America* (1970); Steven Stowe, *Doctoring the South: Southern Physicians and Everyday Medicine in the Mid-Nineteenth Century* (2004); Melbourne Tapper, *In the Blood: Sickle Cell Anemia and the Politics of Race* (1999); Karen Kruse Thomas, *Journal of African American History* (Summer 2003); Keith Wailoo, *Dying in the City of the Blues: Sickle Cell Anemia and the Politics of Race* (2001); Thomas J. Ward Jr., *Black Physicians in the Jim Crow South* (2003); Joseph I. Waring, *A History of Medicine in South Carolina*, 3 vols. (1964–71); Laura R. Woliver, *The Political Geographies of Pregnancy* (2003).

Abortion

Humans look away or deny the evidence of their actual behavior. Southerners are no exception to this rule. Southern behavior belies many dominant images of the region. Although the South is described as the belt buckle of the Bible Belt, that belt is often, in fact, unbuckled. Abortion rates in the South provide one indicator of the complexities of southern reproductive behavior.

Denial about sexual behavior is rampant in the South. For instance, one entire southern state, South Carolina, lived in official denial about the late senator J. Strom Thurmond. Even though many people whispered for decades that Thurmond had an African American daughter, it was only in 2003 that the "open secret" was uncloseted, when Essie Mae Washington made her heritage officially public. What southerners proclaim they abide by and what they actually do regarding sexual behavior are often two different things. Thurmond's story is one illustration.

Given the solid conservative hegemony in southern states currently, it is fascinating to recall that many southern state legislatures were among the first to modernize or liberalize their contraceptive and abortion laws in the 1960s and early 1970s. Although abortion was not the central concern of medical professionals during these decades, according to Gene Burns, in *The Moral Veto*, "the argument that physicians should decide when abortion was appropriate was entirely convincing to a number of state legislatures, especially in the part of the country we would least expect, that is, the South." One puzzle is how the quiet, elite movement to liberalize state abortion laws between 1966 an 1973 was particularly uncontroversial in the South. Women's rights language was not used when liberalizing the abortion laws before the *Roe* decision. Rather than appeal to morality or rights, advocates relied on medical legitimation, permitting medical doctors to be the gatekeepers for legal abortion. Scholars maintain that reform laws, especially when based on humanitarian or medical grounds, were particularly successful in the South during the 1966–73 reform period because no feminist or Catholic groups polarized the issue.

Abortion is a common experience for many women within the United States. Half of all pregnancies in America are unintended and, from those that are unintended, half end in abortion. Approximately 33 percent of American women will have an abortion by the time they are age 45. Abortion rates have recently declined in the United States, though. After the 1973 *Roe v. Wade* decision, abortion numbers continued to rise until 1990 when they reached an all-time high of 1,608,600 abortions per year. Since 1991, abortion numbers declined slowly, to 1,293,000 in 2004.

Since the 1973 *Roe* decision, abortion rates in southern states such as Alabama, Arkansas, Georgia, Kentucky, Louisiana, Mississippi, South Carolina, Tennessee, Texas, and Virginia resemble the national declining trends from 1991 to 2000. However, North Carolina and Florida display increased abortion rates. The reason for disparities between these states and national trends is not known, although it may be because women are seeking abortions in states where clinics are more accessible than in their own states.

There are many possible explanations as to why abortion rates have declined in the United States. Some experts maintain that contraception use is more widespread, while others contend that women are less fertile. Most likely, some decline in abortion rates has been caused by state restrictions placed on abortions. Although the Supreme Court, in *Roe v. Wade* (1973), ruled that a woman has a constitutionally guaranteed right to abort in the early stages of pregnancy, state restrictions have eroded women's right to choose. Conservative southern legislators have been particularly instrumental in limiting abortion since the 1973 *Roe* decision. In fact, abortion rights in the South seem to be in jeopardy.

As can be seen in Table 1, state funding for abortions in Alabama, Arkansas, Florida, Georgia, Kentucky, Louisiana, Mississippi, North Carolina, South Carolina, Tennessee, Texas, and Virginia is given only in cases of emergency such as rape, incest, or life endangerment of the mother. In nine southern states, women must receive state counseling and then wait 24 hours to have an abortion. Eleven states also require that a minor must receive consent from a parent to have an abortion.

Besides state restrictions within the South, partial birth or late term abortion laws are being created to further reduce the number of abortions. In *Stenburg v. Carhart* (2000), the Supreme Court voted against a Nebraska late term abortion ban, stating it was unconstitutional because it did not include any exceptions for the health of the mother. In addition, the Nebraska ban was overly broad. Almost all of the southern states have enacted similar bans on late term or partial birth abortion. Most southern policies on late term abortion are unenforceable because they do not include exceptions for the health of the mother as per the *Stenberg* decision.

Although southern states have officially continued to allow women to have abortions, few clinics exist that can perform the procedure, as can be seen in Table 2. Access to legal abortion throughout the South is dramatically diminishing. Arkansas, Kentucky, and Louisiana have two abortion clinics for each state. South Carolina has three abortion clinics. Mississippi has two abortion clinics within the entire state, down from six in 1996 and four in 2000. North

TABLE 1. *Abortion Laws in Southern States*

State	Parental Consent If Minor[a]	Mandatory Counseling and Waiting Period[b]	No Public Funding Unless Emergency[c]	Not Covered by Insurance[d]	Covered by Private Insurance in Emergency[e]	Public Employees Have Insurance in Emergency[f]	Ban on Partial Birth Abortion
Alabama	*	*	*				*
Arkansas	*	*	*				*
Florida			*				*
Georgia	*	*	*				*
Kentucky	*	*	*	*	*		*
Louisiana	*	*	*				*
Mississippi	*	*	*			*	*
North Carolina	*		*				
South Carolina	*	*	*				*
Tennessee	*	*	*				
Texas	*	*	*				
Virginia	*	*	*			*	*

Source: Alan Guttmacher Institute State Center, 2006.

a. The parent of a minor must consent before an abortion is provided.

b. A woman must receive mandatory state-directed counseling and then fulfill a waiting period before an abortion is provided. Alabama, Georgia, Kentucky, Louisiana, Mississippi, Texas, and Virginia have a 24-hour waiting period. Arkansas and South Carolina have a one-hour waiting period.

c. Public funding is available for abortion only in cases of life endangerment, rape, or incest.

d. Abortion is not covered in insurance policies for public employees.

e. Abortion is covered in private insurance policies only in cases of life endangerment, unless an optional rider is purchased at an additional cost.

f. Abortion is covered in insurance policies for public employees only in cases of life endangerment, rape, incest, or fetal abnormality.

TABLE 2. *Number of Abortion Clinics in Southern States*

State	1996	2000	2004
Alabama	14	14	5
Arkansas	6	7	2
Florida	114	108	26
Georgia	41	26	5
Kentucky	8	3	2
Louisiana	15	13	2
Mississippi	6	4	2
North Carolina	59	55	5
South Carolina	14	10	3
Tennessee	20	16	4
Texas	64	65	22
Virginia	57	46	8

Source: Alan Guttmacher Institute State Center, 2005; National Abortion Federation, 2004.

Carolina had 55 abortion clinics in 2000 and now has five. The numbers of abortion clinics in 2004 continued to decrease.

If a woman wants to have an abortion, she must commute long distances, requiring that she may have to take time off work and have the funds to travel. In addition to the expense of the procedure itself, mandatory counseling, 24-hour waiting periods, and other regulations increase the costs of legal abortion for providers and patients. Relying on a significant other or family member for help with expenses, transportation, or other accommodations may be problematic for a woman who might fear the consequences of seeking such assistance. Additional hindrances to access to legal abortion in the South, as throughout the whole country, are social class, race, immigrant status, and sexuality.

Given that the South is the poorest region in the country, lack of access to financing for southern women adds further hardships to already poor and beleaguered populations. Even before *Roe*, women with money and connections could get their unplanned, unwanted pregnancies taken care of. Abortion restrictions in the South will have a huge social class impact, given the large numbers of southern women without money or safe options.

The future of legal abortion now rests with the United States Supreme Court currently staffed by a majority of justices nominated by Republican presidents and confirmed by a Republican-dominated U.S. Senate. Southern voters were vital determinants of those Republican presidential and Senate victories. Surprisingly, though, when, in 2010, the citizens of Mississippi were faced with Initiative 26—a ballot initiative that would have amended the state constitution to "define the word 'person' or 'persons,' as those terms are used in Article III of the state constitution, to include every human being from the moment of fertilization, cloning, or the functional equivalent thereof"—voters voted down the initiative 55 percent to 45 percent. The amendment would have ended abortion practices in Mississippi altogether and would have set the stage for a new Supreme Court ruling that would have potentially overturned *Roe v. Wade.*

Actually, only some southerners will pay the high cost if abortion is eventually outlawed. (The same groups that bear the brunt of social policies that underfund child welfare programs, understaff social service systems, and underprosecute gender-based domestic violence offenses, to name a few, given their poverty, will also have few safe alternatives to handle unwanted pregnancies.) Policies based on romanticized visions of what southern people need, rather than on their actual, documented, complex behaviors, overlook the demand for affordable, accessible, and safe reproductive health care.

CHRISTINE M. SIXTA
LAURA R. WOLIVER
University of South Carolina

Gene Burns, *The Moral Veto: Framing Contraception, Abortion, and Cultural Pluralism in the United States* (2005); Barbara Hinkson Craig and David M. O'Brien, *Abortion and American Politics* (1993); Donald T. Critchlow, *Phyllis Schlafly and Grassroots Conservatism: A Woman's Crusade* (2005); Karen O'Connor, *No Neutral Ground? Abortion Politics in the Age of Absolutes* (1996); Jean Reith Schroedel, *Is the Fetus a Person? A Comparison of Policies across the Fifty States* (2000); Mary C. Segers and Timothy A. Byrnes, eds., *Abortion Politics in American States* (1995); Rickie Solinger, *Pregnancy and Power: A Short History of Reproductive Politics in America* (2005); Raymond Tatalovich, *The Politics of Abortion in the United States and Canada: A Comparative Study* (1997); Laura R. Woliver, *The Political Geographies of Pregnancy* (2003).

Aerospace

The term aerospace gained currency during the 1950s and was a product of U.S. Air Force nomenclature. It evolved in response to the growing interest of aviation manufacturers and the government in space exploration as well as in traditional aeronautics. From the first flight of the Wright brothers at Kitty Hawk, N.C., to the launch of America's first astronauts from Cape Canaveral, Fla., the South has played an active role in aerospace developments.

On 17 December 1903 the Wrights made the world's first flights in a powered airplane. During the years before World War I, planes were generally viewed as carnival curiosities, although a number of events in the South gave a hint of future trends and made southerners aware of the coming "air age." In 1911 Cal Rodgers completed the first transcontinental flight, which lasted three months. His route took him across Texas, where good weather and level terrain promised an easier course around the Rocky Mountains. Three years later in Florida, promoters launched the country's first commercial airline—a flying boat service across the bay between Tampa and St. Petersburg.

After 1917, when America entered World War I, the attributes of sunny weather and open spaces made the South a center of flight training, a pattern that has persisted across the region. During the 1920s commercial aviation began to expand, especially after the Air Mail Act of 1925, which transferred post office airmail routes to private contractors. Numerous companies launched services, but much consolidation eventually thinned their ranks. On the eve of World War II, major airlines with a southern heritage included National Airlines (1934–91) and Eastern Airlines (1926–91) in Florida, Delta Airlines (1920s–present) in Georgia, and Braniff Airlines (1928–82) in Texas. Also, Pan American World Airways (1927–91) had major routes from Florida into the Caribbean and from Brownsville, Tex., into Latin America.

As southern farmers continued their perennial campaign against the cotton boll weevil, aerial crop dusting grew rapidly during the 1920s. Agricultural aviation owed much to pioneering work conducted by Dr. B. R. Coad and the U.S. Bureau of Entomology in Louisiana. Among the many commercial crop-dusting companies, Huff-Daland of Monroe, La., was one of the most successful, and it eventually grew into Delta Airlines. Aerial crop treatment became even more widespread after World War II; a research-and-development project at Texas A&M during the 1950s produced a unique design with special safety features that became standard for the industry.

During World War II, the South's even terrain and good flying weather again made it a major center for flight training; U.S. Army and U.S. Navy facilities turned out thousands of pilots, navigators, and bombardiers. In the postwar era, many of these training fields continued to serve as operational air bases and as centers of flight training. (As of 2010 the U.S. Air Force Air Education and Training Command is headquartered in San Antonio, Tex., and all of the forces under its command are based in southern states.)

Aviation manufacturers also moved south during World War II to take advantage of climate, available land, and a plentiful supply of labor. The Dallas–Fort

Early rockets exhibited at the Lyndon B. Johnson Space Center in Houston, Tex. (National Aeronautics and Space Administration)

Worth area continued after the war as a major producer of bombers, fighters, and helicopters; plants in Atlanta, Ga., turned out huge transports and other aerospace hardware. Many cities developed facilities for producing electronics and a variety of aerospace products. After the creation of the National Aeronautics and Space Administration (NASA) in 1958, the South became known worldwide for its role in America's space program. The John F. Kennedy Space Center in Florida became NASA's principal launch site; the George C. Marshall Space Flight Center in Alabama played a major role in developing launch vehicles and in handling manned operations; and the Lyndon B. Johnson Space Center near Houston, Tex., was the focal point for astronaut training, mission control, and other tasks. As one historian noted, these centers represented a "fertile crescent" of advanced technology in the South.

These southern NASA centers continue to contribute heavily to NASA's operations. The Marshall Center in Huntsville, Ala., acts as the science command post for the International Space Station and continues to develop new propulsion technology and environmental control and life support systems for people on the International Space Station. Florida's Kennedy Center is still the launch site for every manned mission, including many scientific ones. The Johnson Center in Houston still serves as the lead NASA Mission Control Center for human space programs and houses training facilities for astronauts. Stennis Space Center in south Mississippi is NASA's largest rocket engine test complex. All space shuttle main engines are tested at Stennis, and new launch vehicles, which are expected to replace space shuttles after 2010, are currently being tested there.

The influence of these centers' presences can be seen in the science-technology activities of neighboring schools, universities, and businesses; the centers also create tourism revenue with special visitor programs. Marshall Space Center's presence earned Huntsville, Ala., the name "The Rocket City," and the U.S. Space and Rocket Center there is an extensive museum of space-exploration artifacts, including the Saturn V rocket, and interactive exhibits and simulators. The Space and Rocket Center also hosts Space Camp, Aviation Challenge, and X-Camp programs for the public. The Kennedy Center Visitors' Complex offers opportunities for the public to meet and dine with astronauts, watch launches, and ride in spaceflight simulators. From June to December 2010, the Kennedy Space Center's Astronaut Encounter Theater presented *Star Trek Live*, an interactive production of Star Trek's "Starfleet Academy" for the audience "cadets." Also housed at Kennedy is the U.S. Astronaut Hall of Fame. The revenue from these centers generates substantial income for surrounding communities and provides curriculum fuel and off-campus educational opportunities for area schools.

Commercial air travel expanded in kind in the postwar years, especially after the introduction of jets in the late 1950s. Airport cities became major hubs of national and international significance, and their reach has only grown with time. In 2010 Hartsfield–Jackson Atlanta International Airport was again named the busiest passenger airport in the world, a position the airport has held since 1998; in 2009 it was estimated that the airport handles about 89 million passengers annually. The Dallas–Fort Worth International Airport is, as of 2010, the third-busiest airport in the world, handling about 57 million passengers a year. That airport attracted so many national corporations that the Dallas–Fort Worth area now boasts the heaviest concentration of corporate headquarters in the country, with more than 10,000 corporate headquarters located there.

Until 2010 Delta Airlines, headquartered in Atlanta, was the largest airline in the world, but in September 2010 a merger of Continental Airlines (Houston) and United Airlines (Chicago) superseded Delta's title. The new United/Continental airline is now based in Chicago, so American Airlines (Fort Worth) and Delta (Atlanta) are the only major airlines left in the South.

ROGER E. BILSTEIN
University of Houston at Clear Lake City

MARY AMELIA TAYLOR
Judson College

Roger E. Bilstein, *Flight in America, 1900–1983: From the Wrights to the Astronauts* (1984); Dallas–Forth Worth International Airport, Press Release (17 September 2010); R. E. G. Davies, *Airlines of the United States since 1914* (1972); "Delta No Longer World's Largest Airline," *Atlanta Business Chronicle* (September 2010); Hartsfield–Jackson Atlanta International Airport, Press Release (22 January 2009), "Atlanta Fact Sheet" (2010); W. David Lewis and Wesley P. Newton, *Delta: The History of an Airline* (1979); NASA website, www.nasa.gov (2010); Loyd S. Swenson, *Southwestern Historical Quarterly* (January 1968).

Agriculture, Scientific

Although southerners made remarkable agricultural progress between 1800 and 1860, science contributed comparatively little until the 1870s. Before the Civil War, American agriculturists using empirical methods developed the essentials of modern farming. In the 20th century, science and technology built upon this foundation to make American agriculture the most productive in the world.

In the South, planters started in the 1790s to cultivate new crops of Sea Island cotton, upland cotton, and sugar with slave labor and primitive farming methods based on spades, hoes, and ox-drawn turning plows. Within the span of a single lifetime, they created advanced systems for producing those crops on a vast scale. During the period from 1830 to 1860, they mechanized their farming operations with mule-drawn implements and even used steam engines to power their gins, mills, and presses. By 1850 cotton growers had bred the modern type of upland cotton by crossing varieties imported from the Caribbean islands, Mexico, and Siam and then refining the resultant cotton with selective breeding. During the 1840s, planters adopted horizontal culture of row crops, crop rotation, and elaborate drainage systems in order to preserve the fertility of their farmlands. Finally, they devised effective methods of managing slaves in which rewards replaced threats of punishment. By 1860 southern agriculturists had employed almost all the implements and farming methods still in vogue as late as the 1920s.

In the prewar era, chemistry made several useful contributions to southern agriculture. Edmund Ruffin of Virginia demonstrated that marl could renovate worn-out soils by reducing acidity, and Justus Liebig, the German chemist, analyzed soils to discover that cultivation of crops removed important elements from the soil. From him, southerners learned to plow under the stalks of their cotton and corn. Liebig's research also pointed the way to improving soil fertility by adding missing elements.

Many of the South's agricultural achievements were lost as a result of the Civil War. Plantations were subdivided into family-sized farms worked by unsupervised sharecroppers. In the process, the economies of large-scale farming disappeared, and the improved management techniques of the 1850s became inapplicable. The prewar trend toward farm mechanization ended, and soil-conservation systems were abandoned. Inevitably, the productivity of farmlands and agricultural workers diminished. Had it not been for two contributions made at this time by chemistry, the South's agricultural economy would have declined still further. Newly introduced commercial chemical fertilizers partially offset the loss of soil fertility characteristic of sharecropping, and the first arsenic-based insecticides reduced crop losses.

Around the turn of the 20th century, state and federal governments established experiment stations employing chemists, botanists, and entomologists to seek solutions for agricultural problems. In the private sector of the economy, commercial plant breeders who earlier had relied on selective breeding now began to apply the science of botany with noticeable effect.

Between the two world wars, gasoline tractors benefited southern agriculture even more than improvements made in chemical fertilizers and insecticides. Early models assumed the heavy labor of breaking land for planting, and later tricycle types increasingly performed much of the cultivation formerly done with mule-drawn implements. With the coming of tractor-drawn, multirow mechanical implements, landowners regained much authority over farm workers that had been lost with the collapse of slavery. A new system of day labor began slowly to replace sharecropping. By 1940 mechanization had progressed so far that crops of cotton could be planted and cultivated almost entirely with machines. Manual labor was required only for a small amount of hoeing and for harvesting the crop.

Between 1945 and 1950, the mechanization of southern agriculture was completed. Mechanical cotton strippers and cotton pickers now harvested the South's principal crop, and flame-throwing devices and chemical herbicides replaced the last of the hoes. With the advent of these revolutionary machines, the system of sharecropping became obsolete, and by 1960 it was virtually dead.

In the place of sharecropping arose a new system of large landholdings worked as single units, not unlike the old slave plantations, with diesel- and gasoline-powered machines taking the places of slaves and mules. With the emergence of consolidated mechanized plantations in the 1960s, the age of scientific agriculture finally dawned in the South, decades later than in other sections of the nation. Landowners who had gained full control over their farming operations began to effectively apply modern techniques of management, introducing the new products of science and technology into their system of agriculture. Machines distributed a new generation of chemical fertilizers, insecticides, fungicides, and herbicides with scientific accuracy. Scientific plant breeders supplied farmers with prolific new plants tailored for disease resistance and ease of harvesting by machine. In all major crops, yields per acre increased dramatically.

In the post–World War II era, southern agriculturists acquired a new versatility from science and technology. They were able to contend with changes in the market by shifting from cotton to supplementary crops of soybeans or small grains. In some areas, cotton was entirely abandoned for rice, corn, or peanuts.

Scientific and technological progress in southern agriculture brought new problems as well as benefits. Millions of farm workers lost their employment, and rural populations declined. With mechanization, the South became dangerously dependent on the international oil industry for fuel and agricultural chemicals. Insect pests demonstrated a dismaying capability of developing resistance to insecticides, and hybrid plants proved vulnerable to epidemic viral plant diseases. Thus far, chemists have overcome insect resistance with new toxic substances, but the future of this approach is clouded because the number of suitable chemical combinations is limited. Agricultural scientists therefore are turning to the control of insects through both natural enemies and sterilization with radiation.

Twenty-first-century advances in modern science have greatly affected agricultural practices across the region, particularly in the field of biotechnology. Agricultural biotechnology, the practice of genetically engineering crop plants to tolerate specific herbicides and resist specific plant diseases and insect pests, has seen its share of controversy since the first crop varieties developed by genetic engineering were introduced for commercial production in 1996. The benefits of biotechnology to farmers—particularly corn, cotton, and soybean farmers in the United States—are significant. For instance, plants that tolerate certain herbicides allow farmers to effectively spray fields for weeds without damaging crops, which improves plant quality and often produces larger crop yields. Also, crops that resist certain insect pests allow for the reduction in use of environment-threatening synthetic pesticides, and crops that resist various diseases also tend to produce higher yields.

According to the Pew Initiative on Food and Biotechnology, 97 percent of Mississippi's 1.1 million acres of cotton planted in 2004 was genetically modified, followed by 94 percent of Georgia's 1.33 million acres, and 94 percent of Arkansas's 950,000 acres. In terms of acreage in production, Texas led the nation in genetically modified cotton in 2004, with 58 percent of its 6 million acres of cotton planted (up from 49 percent in 2001). For soybeans, Mississippi and Arkansas ranked second (93 percent) and third (92 percent), respectively, in percentage of acreage planted in 2004, following only South Dakota (95 percent). In 2004 the South planted relatively little genetically modified corn as compared to the 11 top-producing states, all located in the Midwest and Upper Midwest.

Controversy lies in so-called Frankenstein foods. Opponents of genetically modified foods question the safety of long-term consumption of these products by humans and livestock. Little to no long-term research has been conducted, although many crops for human consumption and forage have been approved by the Food and Drug Administration and the Environmental Protection Agency. Questions also abound regarding the environmental safety of the crops. Partially in response to these concerns, an organic foods movement has gained momentum across the South and entire United States, with farmers producing organic foods at levels not seen since before widespread use of chemical fertilizers and insecticides and making the organic foods market the fastest-growing sector in the U.S. food industry.

JOHN HEBRON MOORE
Florida State University

JAMES G. THOMAS JR.
University of Mississippi

Gilbert C. Fite, *Agricultural History* (January 1979, January 1980); John L. Fulmer, *Agricultural Progress in the Cotton Belt since 1920* (1950); Paul W. Gates, *The Farmer's Age: Agriculture, 1815–1860* (1960); Douglas Helms, *Agricultural History* (January 1979, January 1980); Willard Range, *A Century of Georgia Agriculture, 1850–1950* (1954); Charles R. Sayre, *Agricultural History* (January 1979); Richard C. Sheridan, *Agricultural History* (January 1979); Michael R. Taylor, Jody S. Tick, and Diane M. Sherma, *Tending the Fields: State and Federal Roles in the Oversight of Genetically Modified Crops* (December 2004); U.S. Department of Agriculture, "Opportunities and Challenges in Agricultural Biotechnology: The Decade Ahead; A Report Prepared by the USDA Advisory Committee on Biotechnology and 21st-Century Agriculture" (13 July 2006), "Preparing for the Future: A Report Prepared by the USDA Advisory Committee on Biotechnology and 21st-Century Agriculture" (9 May 2005).

Alcohol and Alcoholism

"Ambivalence," the label social scientists most often apply to American attitudes toward alcohol, fits southerners particularly well. Extremes of opinion and practice can be found in practically any southern community, with teetotalers condemning alcohol and good old boys swilling it in manly ritual. Will Rogers joked that some southerners did both, voting dry "as long as they can stagger to the polls."

It was not always so. Southern colonists and their descendants in the early years of the Republic had few qualms about alcohol; they drank hard and often. The Virginia Company was plagued by planters who crowded aboard floating taverns in the James River, bartering their tobacco for spirits and sack. They, like later frontiersmen, caroused to escape the loneliness and hardship of wilderness life. Even in well-established towns and plantations, however, drinking was nearly universal among adult white males. Men drank upon arising and retiring, during and between meals, and while celebrating holidays, recuperating from illness, conducting business, and soliciting or pledging votes.

Southerners drank a variety of beverages. They distilled brandy from apples, peaches, pears, and other local fruit. Apple cider was also popular, if not as ubiquitous as in the North. Imported wines, especially Madeira and claret, graced wealthy planters' tables. All classes drank rum, obtained in exchange for southern commodities. Rum fortified most strong drinks of the 18th century, including punch, flip, toddy, grog, and blackstrap. It also served as an important, if controversial, means of bartering with the Cherokees and other southern Indian tribes.

When the Revolution disrupted trade, making supplies of West Indian molasses expensive and uncertain, rum consumption declined. Its place was taken by whiskey, a drink familiar to Finnish, Swedish, Scottish, Irish, and Scots-Irish immigrants. Their knowledge of making whiskey, combined with improved stills and ample grain, water, and fuel, assured an abundant supply. Distilling was especially important in the corn-growing areas of the Upper South—the word "bourbon" derives from Bourbon County, Ky. Southerners such as Thomas Jefferson, alarmed by the deluge of cheap spirits, proposed viticulture and brewing as more salubrious alternatives. However, domestic wines and beer did not make significant inroads against whiskey in the South until the mid-20th century.

Antebellum southerners paid for their indulgence with high rates of alcoholism and violence. The violence arose not only from drinking but from drinking in an environment where weapons were ubiquitous and men were homicidally sensitive about honor—Colonel Sherburn's motive for slaying the drunken Boggs in Mark Twain's *The Adventures of Huckleberry Finn* (1885). Slaves suffered as well, for there was little they could do to protect themselves from a master turned brutal by drink. Harriet Beecher Stowe capitalized on their plight in *Uncle Tom's Cabin* (1852), which contains several pointed references to Simon Legree's drinking.

The temperance movement in the antebellum South was relatively weak. Although it had 44 percent of the population, the South accounted for only 8 percent of the nation's temperance pledges in 1811. No slave state, save Delaware, had adopted prohibition by the 1850s. A perceived link with antislavery hurt the temperance cause in the South, as did the economic circumstances of isolated farmers, who depended on distilling to retard spoilage, reduce bulk, and enhance the marketability of their crops. Some of the more substantial farmers and aspiring middle-class townsfolk joined temperance societies, but their influence was outweighed by the planter elites, who remained aloof—and conspicuously wet.

The Civil War had mixed consequences for southern drinking. Temperance societies were disrupted during the war, and defeat gave many demoralized southerners added cause to resort to the bottle. Yet, in other ways, the war paved the way for the eventual triumph of the drys. The war set a precedent for prohibition; during 1862 Confederate legislatures sought to preserve grain by outlawing its distillation. Whiskey prices rose sharply. They fell again after 1865, but not to antebellum levels, because of the retention of federal excise taxes on beer and liquor. Bootleggers, of course, did not pay taxes, but trouble and risk necessarily inflated the cost of their product. The long-term effect of higher prices was to discourage consumption.

In destroying planter hegemony, the war also made possible the political rise of the middle class, which was much more hostile toward drink. Middle-class reformers were quick to climb aboard the prohibition bandwagon, denouncing alcohol in the name of economy, discipline, honest government, and other progressive virtues. Populists also hated the liquor dealers, whom they accused of exploiting the people and, not without evidence, of manipulating their representatives. They were joined on this issue by many New South industrialists, who saw liquor as undermining productivity. Evangelical Protestants, whose numbers and influence grew rapidly in the late 19th century, supported the dry

Paul Newman and Elizabeth Taylor as Brick and Maggie in Cat on a Hot Tin Roof *(1958), in which alcoholism was a prominent theme (Film Stills Archives, Museum of Modern Art, New York)*

alliance. Ministers admonished churchgoers, chided backsliders, and vocally supported antiliquor legislation.

Conservatives who distrusted government meddling and who thought that what a man drank was his own business opposed the dry crusaders. So did urban political machines and the still-formidable liquor interests. The best the wet coalition could manage, however, was a delaying action against prohibition. Its natural allies, immigrants with cultural backgrounds favorable to drinking, had mostly settled outside of the South. Wets courted native-born, lower-class voters of both races as an alternative source of support, but, as disfranchisement thinned their ranks, it became harder to stem the prohibitionist tide. Local-option elections and special legislation dried up more and more territory. By 1907, 825 of the 994 ex-Confederate counties had some form of prohibition. From 1907 to 1909, drys won a round of statewide victories, as Georgia, Oklahoma, Alabama, Mississippi, North Carolina, and Tennessee all enacted statutory prohibition.

Prohibitionists acted from religious, political, and humanitarian motives consistent with the progressive spirit of social uplift. But they were also plainly interested in social control, especially of blacks. Before the Civil War, law and custom confined plantation slaves to an occasional holiday spree. Emancipation loosened these restraints. Prohibitionists exploited this situation by alleging that atrocities were committed by drunken blacks; "nigger gin" joined "demon rum" as a favored epithet. D. W. Griffith's *The Birth of a Nation* (1915) gave cinematic expression to these fears. Drawing upon two earlier novels by Thomas Dixon Jr., Griffith portrayed freedmen who were drunken, arrogant, and lecherous, in contrast to their sober, docile, and hardworking slave forebears. Not to be outdone, some wets played up stories of cocaine rampages—the implication being that, if blacks could not drink, they would turn to more dangerous drugs.

Dry propaganda aside, postbellum blacks did not have a serious alcohol problem. On the contrary, they drank less than poor whites, especially in rural areas. But the situation changed during the 20th century, as uprooted blacks began migrating to cities, where morals were looser, liquor was more abundant, and good, steady jobs were scarce. Poor black males living in cities (southern or otherwise) were more likely to develop drinking problems, and to develop them sooner, than either the general population or their country relations. That was why, for example, the District of Columbia, with its large black ghetto, had exceptionally high rates of alcoholism and liver cirrhosis deaths in the late 20th century. Yet Alabama and Mississippi, two states with large rural black populations, had unusually low rates during the same period. To generalize about southern blacks and alcohol, one has first to be specific about place. Local conditions mattered.

The same could be said of the South generally. Since World War II, the South, as a region, has ranked at or near the bottom of national rates of alcoholism and per capita alcohol consumption. The percentage of southerners who abstained from alcohol also remained relatively high. The usual explanation is that the South remained disproportionately Protestant, rural, and dry, as many areas retained prohibition long after national repeal. Yet that was not true everywhere. Throughout the late 20th century and into the early 21st century, Louisiana and Florida maintained high levels of per capita alcohol consumption relative to both the South and the country as a whole. These two states had proportionately more Catholics, more immigrants, and more city dwellers than most other southern states. Louisiana also had New Orleans, a city long associated with alcoholic indulgence. Like the bourbon-fueled Texas wildcatters or the mint julep crowd at the Kentucky Derby, the revelers of New Orleans's French Quarter came to symbolize the excesses of "southern" drinking—excesses that lived on in the popular imagination, if not in sober statistical facts, about the region's average alcohol consumption.

DAVID T. COURTWRIGHT
University of North Florida

Joe L. Coker, *Liquor in the Land of the Lost Cause: Southern White Evangelicals and the Prohibition Movement* (2007); N. E. Lakins et al., *Apparent Per Capita Alcohol Consumption: National, State, and Regional Trends, 1977–2006* (2008); C. C. Pearson and J. Edwin Hendricks, *Liquor and Anti-Liquor in Virginia, 1619–1919* (1967); W. J. Rorabaugh, *The Alcoholic Republic: An American Tradition* (1979); Lori Rotskoff, *Love on the Rocks: Men, Women, and Alcohol in Post–World War II America* (2002); James B. Sellers, *The Prohibition Movement in Alabama, 1702–1943* (1943); Muriel W. Sterne, in *Alcoholism*, ed. David J. Pittman (1967); Joe Gray Taylor, *Eating, Drinking, and Visiting in the South* (1982); Ian R. Tyrell, *Journal of Southern History* (November 1982); Daniel Jay Whitener, *Prohibition in North Carolina, 1715–1945* (1945).

American Indian Health and Medicine

Early European travelers found the inhabitants of the Western Hemisphere to have developed extensive knowledge and efficient use of their natural resources. Archaeology is uncovering evidence that the South could have been inhabited for at least 16,000 years, as shown at the Topper Site in the central Savannah River area of South Carolina, with the oldest radiocarbon dated material in the country. It is no mystery, then, that southern tribes have had many generations to systematically study and utilize their diverse landscape and its contents. With the collective knowledge of flora, fauna, and ecosystems, tribes systematically observed, measured, and experimented with plant knowledge and other resources to develop a sophisticated and effective pharmacopeia. Early travelers among southeastern tribes such as John Lawson noted the use of various plants and returned to London with many specimens for pharmaceutical study. Once knowledge of medicinal plants was shared with newcomers, large-scale harvests of plants like ginseng and sassafras were exported to European and Chinese markets in quantities enough to fund a multitude of voyages. Plants from the New World became a major economic commodity in a world trade market by the mid-18th century.

The health of southeastern Indians at Contact (AD 1500) varied then, as do populations today. Depending on diet, lifestyle, and location, Native peoples suffered mainly from three classifications of disease, as physical anthropologists Verano and Ubelaker have pointed out. The first classification was the infectious diseases, caused by bacteria, viruses, parasites, or fungi. For example, parasites including tapeworms, hookworms, roundworms, whipworms, and pinworms were relatively common. Evidence of disease from skeletal material usually indicates the presence of chronic infectious bacterial illnesses such as the four syndromes associated with treponematosis and other infections. Anemia was also problematic in some populations and appears in the skeletal material from those groups. Tumors indicative of various cancers are also apparent. The second classification included traumatic lesions, which were mostly injury related, and the third classification was disease of the joints related to use and stress, such as arthritis and rheumatoid arthritis.

Virgin soil epidemics were multiple instances of decimation through disease related to initial contact with new European immigrating populations. Domestic animal–derived diseases, taken for granted as part of childhood among Europeans (such as measles, mumps, and chicken pox), found Native populations who were not physiologically, psychologically, or culturally prepared for their effects, which extinguished whole tribes and reduced the Native populations by more than 90 percent. Historian Peter Wood estimates that the population of Southeast Indians was reduced to a staggering 3 percent of the total population of the South by 1790. New pathogens, along with the brutality of colonization, took its toll on American Indian people. The experiences and effects of early contact are best exemplified through the writings of Bartolomé de Las Casas, an early Spanish settler who became a Dominican friar after his experience. He is considered to be the first activist for human rights in the New World. He chronicles in ghastly detail the inhumane treatment and murder of the indigenous peoples at the hands of Spaniards in the early 1500s. Through records such as his, one begins to understand what little consideration of humane and respectful treatment most early European travelers had for the inhabitants of the Western Hemisphere. European attitudes and paradigms of "Indians as savages" translated into catastrophic health problems for Native peoples of the South.

David S. Jones, in his important contribution to this topic, *Rationalizing Epidemics*, discusses the clash of beliefs and biology between the early colonists and Native peoples. He speaks of the plagues and epidemics that eliminated 90 to 95 percent of Native populations in the early 1600s, with smallpox hitting hard those tribes who met the earliest colonists. He notes that disease is enveloped in larger social, political, and economic contexts that have direct impact on how disease is approached and understood. It is during this time that clerics such as John Winthrop marveled at how God had dealt with the savages to provide a land cleared for use and settlement by colonists. It is no wonder that the development of policy aimed at elimination and ultimately assimilation of Native peoples would work to also legitimize their health disparities. Eurocentric, paternalistic, and Christian-centered paradigms rationalized the often harsh and traumatic policies forced upon American Indians. Rationalization of destroyed towns and crops, forced removal from homelands, separation of families from their children and grandchildren, and attempted obliteration of their language and spiritual beliefs have made an impact, not only culturally and economically, but in significant health disparities.

By chronicling types of health issues since 1600, Jones further illustrates the plethora of health problems experienced by American Indians. Once initial disease clusters of smallpox, influenza, and measles reduced Native populations by 90 percent, federal Indian policy, which was handled by the War Department, created situations that concentrated and confined Native peoples to reservations and institutions like boarding schools. As a result, tuberculosis became a primary health problem until well into the 20th century. Poor health related to poverty, such as malnutrition and communicable diseases, continued to plague Indian communities with the nation's highest infant mortality rates and rates of infectious diseases such as trachoma, which was as high as 95 percent in some Indian schools. In 1921 Congress passed the Snyder Act providing United States citizenship to American Indians but also identifying "relief of distress and conservation of health of Indians." The Merium Survey in 1928 reported the severe conditions in which Native peoples lived and gave impetus for the Indian Reorganization Act of 1934 (otherwise known as the Wheeler-Howard Act), which allowed the federal government to become trustees of Native lands and funds and tribes to establish their own governments and buy back lands that had been taken from them after the Dawes Act of 1887. Retaining and extending the tribal land base were important parts of reestablishing tribal government and direction of their own affairs. In 1955 Indian health was transferred to the Department of Health, Education, and Welfare, now the Department of Health and Human Services, after several studies revealed the continued poor health and living conditions of our country's Native people.

Today Native peoples are working diligently to deal with what researchers Jennie Joe and Robert Young have referred to as the "diseases of civilization"—diabetes, heart disease, and substance abuse. These noncommunicable diseases are the latest health and social challenges facing Native peoples today.

Native Healing Systems. Native populations, like all other human populations, have had to develop healing systems to deal with disease. The insurmountable and rapid rate of mortality experienced by southeastern tribes during the first several centuries of contact affected these systems by eliminating many who cared for the ill and by the loss of knowledge these men and women had kept and passed down for centuries. With that said, the existence of those tribal members who continue to use their medicines and ritual is a testimony to the value of the knowledge they have and the principles of the science they use.

There is good documentation of Indian medicine as recorded by early travelers, naturalists, and historians among southeastern tribes. In the early 1800s there were some who took advantage of their interactions with tribes and created "patented medicines" marketed in the form of tonics, liniments, and droughts. These "medicines" were often marketed as cure-alls and touted American Indian herbal ingredients. One of the most enduring of these is S.S.S. Tonic, which was reportedly a Creek Indian formula that had been given to Captain Dennard of Georgia for saving the life of a chief. He in turn sold the formula to Col. Charles Swift, who, by the late 1800s, was one of the most successful businessmen in the South. But most of these "medicines" had no association with tribes, nor had they been created with a formula used or endorsed by tribes.

Not until recent years do we have contributions of Native people in the discussion of Indian medicine and health. Early historic records and observations by missionaries, naturalists, and others fail to include discussion of how use of ethnobotanicals and administration of care were and are decidedly different from Western medicine.

Native science pioneer Gregory Cajete writes, "Native American science is incomprehensible to most Westerners because it operates from a different paradigm. Measurement is part of Native American science, but does not play the foundational role that it plays in Western science. Measurement is only one of many factors to be considered." He goes on to say that the "Native American paradigm is comprised of and includes ideas of constant motion and flux, existence consisting of energy waves, interrelationships, all things being animate, space/place, renewal, and all things being imbued with spirit." As this works in medicine and healing, ethnobiology is important, but it often cannot work without understanding the larger context in which it is administered. The emphasis is on the dynamic of relationship between patient and provider. In traditional healing, the identification of illness may not be of issue until several visits have passed. The medicine person must first be "centered," or in the right balance, to treat an individual. The medicine is created specifically for that individual, depending on time of month, year, and season and on the age, sex, and other factors relating to the patient. One can read about use of plants, animals, directions, medicine wheels, hot houses, and other tribally unique elements used in Indian medicine, but the substance that brings them all together to work for each individual is the belief that we all are made of and are part of a larger universal system with related energy and elements. This is manifest through tribally specific ritual and language.

Many Southeast Indian people have felt disenfranchised by a health-care system that has been institutionalized and operated on an exclusive Western medical model. Some of these health-care systems have been more recently taken back by tribal contracts with the government, but the Indian Health Service is still a very real presence. The Indian Health Service has provided tribal people with improved health care since its early mandate in the 1800s, but there are still serious health disparities.

Native resiliency has fostered a growing need for cultural competency and a revamping of health-care needs with the input, voices, and perspectives of Native peoples. With that comes a necessity for health-care professionals to learn more about Native culture, language, and history.

LISA J. LEFLER
THOMAS N. BELT
Western Carolina University

William L. Anderson, Jane L. Brown, and Anne F. Rogers, *The Payne-Butrick Papers* (2010); Gregory Cajete, *Native Science: Natural Laws of Interdependence* (1999); David Dary, *Frontier Medicine: From the Atlantic to the Pacific, 1492–1941* (2008); David H. DeJong, *If You Knew the Conditions: A Chronicle of the Indian Medical Service and American Indian Health Care, 1908–1955* (2008); Albert C. Goodyear, in *Paleoamerican Origins: Beyond Clovis*, ed. Robson Bonnichsen, Bradley T. Lepper, Dennis Stanford, and Michael R. Waters (2005); Jennie R. Joe and Robert S. Young, *Diabetes as a Disease of Civilization: The Impact of Culture Change on Indigenous People* (1994); David S. Jones, *Rationalizing Epidemics: Meanings and Uses of American Indian Mortality since 1600* (2004); Bartolomé de Las Casas, *In Defense of the Indians*, trans. Stafford Poole (1992); John Lawson, *Lawson's History of North Carolina* (1714, 1937); Everett R. Rhoades, *American Indian Health: Innovations in Health Care, Promotion, and Policy* (2000); Robert B. Shaw, *History of the Comstock Patent Medicine Business and Dr. Morese's Indian Root Pills* (1972); John W. Verano and Douglas H. Ubelaker, eds., *Disease and Demography in the Americas* (1992); Peter H. Wood, in *Powhatan's Mantle: Indians in the Colonial Southeast*, ed. Gregory A. Waselkov, Peter H. Wood, and Tom Hatley (2006).

Childbirth, Antebellum

Childbearing was the central life experience for nearly all southern women during the colonial and antebellum periods. Because most women married or lived with a male partner and because birth control devices were unknown or of limited effectiveness, a woman in good health could anticipate a pregnancy every two to three years during her fertile years. Society glorified motherhood as woman's sacred occupation.

Few details are known about childbirth during the colonial period. Research on the 17th-century Chesapeake region reveals that women there tended to bear children later than the norm because a significant number of young women immigrated to this country as indentured servants. Having to fulfill their indenture meant that they usually married late and did not bear their first child until their mid- to late 20s. The unhealthy southern environment created a tentative situation for newcomers but especially for childbearing women; miscarriages and infant and maternal deaths were common.

Through the end of the 18th century, most black and white southern women depended on midwives or female kin to assist with delivery. Giving birth was usually an all-female experience, with several women assisting a parturient woman with what could be a prolonged delivery. Nearly all women delivered their babies at home. Midwives and female attendants let nature take its course. Other than administering herbal concoctions and keeping up a woman's spirits, they did little to interfere with what was regarded as a normal process. Women had few effective means to limit their number of pregnancies, though, in the early years of colonial settlement, miscarriages and high infant mortality kept a check on large families. As the number of indentured servants decreased, families stabilized, and settlers grew more accustomed to the southern environment, women began to bear more children. As the first federal census of 1790 showed, the new nation was surprisingly fertile. Women of childbearing age averaged about seven live children.

A woman's primary caretaker during childbirth was her mother, and some southern women made every effort to return home to their parents for their delivery. Female attendants continued to be important in the birthing room. Husbands rarely played a major role other than to deliver news of the newborn. Few women went to a hospital to have a baby, for hospitals were few in number and generally regarded as unsafe places where doctors practiced their skills on the urban poor, who constituted the majority of patients.

Changes in childbearing occurred in the antebellum period. The process became more professional as male physicians became involved in the birthing process, most regarding it as a pathological rather than a natural condition.

More men entered the medical profession and gained requisite skills by being apprenticed to an older doctor or attending one of the many proprietary medical schools established in the North and South. Though medical education proved of limited value because of a rudimentary understanding of science and medicine at the time, obstetrics became a popular field. To enhance their reputation, physicians denigrated midwives as old-fashioned and unskilled. Several wrote advice manuals to provide information on pregnancy and childbirth. Doctors who were seeking respectability and greater income found their role as a birth attendant an effective means to gain an entire family as patients. More physicians began to consider themselves experts in this field, whether that reputation was deserved or not. Because status surrounded the use of a male doctor in childbirth, more privileged southern white women began to depend on a doctor to deliver their babies.

Despite education and training, doctors had limited medical understanding of the birth process. They did not yet have a clear idea of the process of conception and believed a woman was most likely to get pregnant immediately before or after menstruation. Education on childbirth in medical school was done by lecture rather than live demonstration in order to preserve a woman's delicacy. Physicians who engaged in the birthing process did so with curtains and bedclothes drawn in order not to embarrass the parturient woman. In contrast to midwives, doctors were more likely to intervene, using their hands, drugs, bloodletting, and instruments to ease or hasten delivery. Taking a lead from Europe, some physicians used instruments such as forceps, tongs, and hooks for difficult births to help dislodge the fetus in order to save the mother's life. Many southern doctors believed in heroic medicine and often administered drugs and bled a parturient woman, believing that dramatic (heroic) means would reduce pressure and ease the baby's birth. The sense among physicians was that the unhealthy southern environment demanded dramatic techniques. These could produce negative results, for the importance of sepsis and an understanding of the harmful effect of many drugs were not yet understood. A doctor's unwashed hands, clothing, and instruments could foster infection. Although outbreaks of puerperal fever usually occurred in hospitals, even rural southern women fell victim to this deadly, infectious disease.

Surgery began to play a minor role in childbirth. Dr. J. Marion Sims in Alabama made a medical breakthrough in 1849 by discovering a surgical cure for vesicovaginal fistula. A tear in a woman's birth canal during childbirth could lead to a lifetime of discomfort, seepage, and embarrassment. Without using anesthesia, Sims performed repeated surgeries on slave women to refine this surgical technique. Until the Civil War, no southern doctor had found a successful means to perform a Caesarean, although a few experimented with it in emergency situations.

Though more women began to depend on doctors, there is little indication that childbirth became safer, although in a few cases it became less painful. A few doctors began to administer ether or chloroform to the parturient woman, with varied results. Women found that anesthesia lessened or eliminated pain, and personal accounts by a few women who used it were positive. Many doctors, however, remained unconvinced of its usefulness. For one thing, they could no longer determine the depth or location of a woman's pain to help guide them through the process. Others believed in the biblical dictate "in sorrow thou shalt bring forth children." If a woman did not experience pain, so the thinking went, she might not adequately love her child. Moreover, many doctors were not trained to know how to use the proper amount of anesthesia, and an overdose could foster dire results.

Few antebellum women publicly announced their pregnancy, but if they did so, it was in a circumspect manner. Medical advice books now offered women information on how to conduct themselves during pregnancy, including what to eat and how to exercise. Traditional tales shared among women cautioned them to monitor their behavior, because many believed that any action or "maternal impression" during pregnancy could have lifetime implications on a child. Few antebellum women sought prenatal care from a doctor unless they were truly ill. Although pregnancy could be an uncomfortable and often unhealthy experience, most women tried to carry on as normal, tending to their myriad domestic activities, attending church, visiting family and friends, and raising their children. Most had little choice. Only the most privileged could indulge by luxuriating in an experience that was viewed as an illness rather than a normal event.

Most women spent time during each pregnancy in a state of fear over what lay ahead. They had good reason, because a number of women died during or right after childbirth. The 1850 federal census shows that nearly twice as many southern women died in childbirth as women in the Northeast. Various reasons help explain this difference. Bearing so many children in the unhealthy southern disease environment took its toll on women's health. Malaria, endemic to the South, had a negative effect on women during pregnancy—fostering miscarriages and debilitation and increasing women's susceptibility to anemia and other diseases. The high fever accompanying malaria could prove fatal to the fetus.

A few couples in the urban Northeast were beginning to practice some form of birth control, but little evidence indicates that many southern couples consciously tried to limit the number of babies a woman bore. Southern families tended to be large, and it was not unusual for a woman to bear a dozen children. Breastfeeding could delay menstruation for up to a year, and abstinence or absence proved effective in postponing conception. Rural couples valued large families, for each child born meant another potential laborer and higher farm productivity. Southern men generally felt pride in a large family, seeing their many children as a reflection of masculinity and honor, though with little regard to how bearing so many children affected their wives' health.

Despite the growing importance of male doctors in childbirth, many southern women still relied on black or white female attendants. They also depended on their deep sense of faith to get them through each delivery, and they invariably thanked God for a successful outcome.

Slave women were valued not only for the work they performed but for their childbearing ability, because each slave child born increased a plantation owner's wealth. For their attendant, slave women usually depended on a midwife in the slave community, although a slave owner might call in a physician if the parturient woman was in distress. Most white southerners, including many doctors, assumed that hard-working slave women had an easier time in childbirth than did white women. Owners encouraged the birth of many slave children and often rewarded the mother with time off from work, lighter duties before and after delivery, or perhaps a small gift. Unlike white women, who were taught to shun sex before marriage, it was not unusual for a slave woman to bear a child before she married. The slave community attached no stigma to premarital sex. There is some evidence of greater spacing of babies born to slave women, who gave birth every two and a half to three years. This may have reflected delayed weaning, a poor diet, bad health, or the absence of a husband or male partner. Evident to all observers was the skin lightening of the slave population, especially in the Upper South. Some white men engaged in interracial sex with slave women, either forced or consensual. Infant mortality was high in slave quarters, because of limited time for maternal care, poor diet, filthy living conditions, and infections and diseases such as neonatal tetanus.

Despite the risks involved and the possibility of ill health or even death, southern women endured many births and treasured each baby born. They prayed that each child would be healthy and reach adulthood. Southern women embraced their sacred occupation with amazing fortitude and commitment.

SALLY G. MCMILLEN
Davidson College

Lois G. Carr and Lorena Walsh, *William and Mary Quarterly* (October 1977); Judith Walzer Leavitt, *Brought to Bed: Childbearing in America, 1750–1950* (1986); Sally G. McMillen, *Motherhood in the Old South: Pregnancy, Childbirth, and Infant Rearing* (1990); John Harley Warner, in *Sickness and Health*, ed. Judith Walzer Leavitt and Ronald L. Numbers (1985); Deborah Gray White, *Ar'n't I a Woman? Female Slaves in the Plantation South* (1986).

Civil War Medicine

The difficulty in determining the exact number of physicians serving in the Confederate Medical Department—and virtually everything else related to medicine in the South during the war period, from mortality rates to numbers hospitalized—is exacerbated by the fact that a considerable portion of the Confederate States' records was destroyed in the burning of Richmond at war's end. Confederate surgeon Joseph Jones's estimate has been questioned, but it may be the best we have: he put the total number of medical officers in the Confederate army at 834 surgeons, 1,668 assistant surgeons, with 73 more in the Confederate navy. Unlike the Union Medical Department, initially bogged down with old hangers on, the Confederate medical corps had no traditions placing its officers in an administrative straightjacket. Thus, while the challenge of building an efficient and effective medical corps from scratch would be formidable, the South was able to start unencumbered.

It was within this context that Samuel Preston Moore was (after the brief tenure of Dr. David C. DeLeon) appointed surgeon general of the Confederacy on 31 July 1861. Moore could be overbearing, officious, and obsessive over details, regulations, and protocols. But at the same time he had all those attributes essential to an able administrator: he was focused, methodical, confident, innovative, and doggedly determined in the face of adversity. With Moore at the helm, an administrative structure was established to provide and distribute medicines to large and widely disbursed forces copied largely from that of the North yet without its ossified adherence to inefficient traditions. Despite poor transportation, a blockade of every major southern port that placed a stranglehold on foreign imports, and an inflationary spiral in medical supply prices, the Confederacy attempted to meet these formidable challenges with improvisation and resolve. At other times, however, the administrative structure either impeded the process of getting medicines to the troops or simply broke down under numerous, insurmountable strains.

An essential element in understanding the nature of medicine during the Civil War in general is the appreciation of the fact that disease was far and away the greatest problem for the militaries of both sides of the conflict. Dysen-

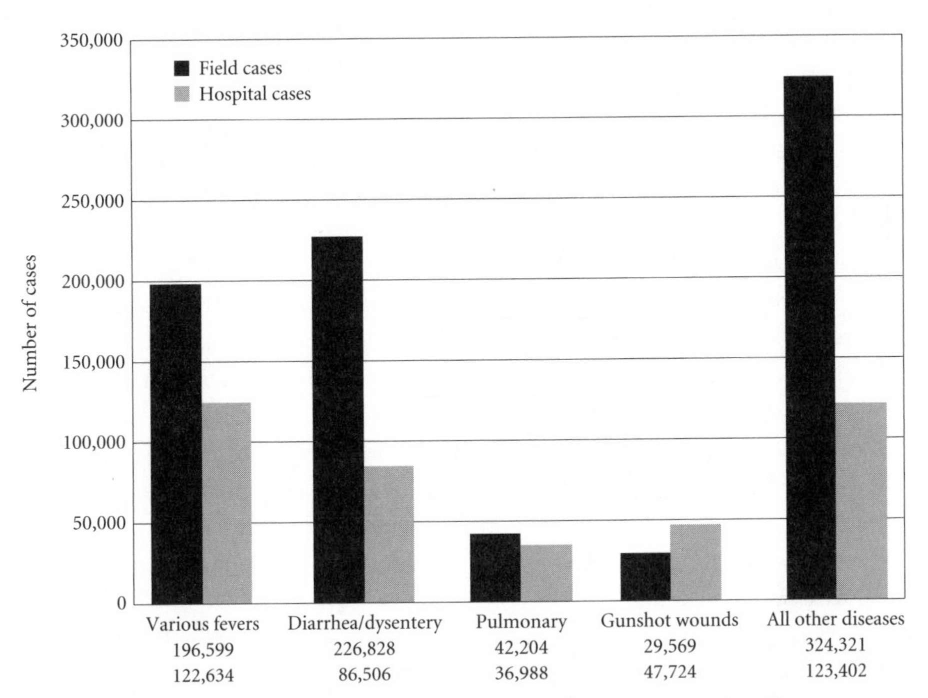

FIGURE 1. *Confederate sick and wounded: Reports on file in the Surgeon General's Office, 1861–1862.*
Source: *"Grand Summary of the Sick and Wounded of the Confederate States Army under Treatment during the Years 1861 and 1862,"* Confederate States Medical and Surgical Journal 1, *no. 9 (September 1864):* 139–40.

tery was the number 1 killer with various camp fevers also devastatingly persistent. Out of more than 6 million medical cases treated (Union and Confederate), only about 400,000 were battle-related injuries. In general, soldiers for the Union *and* the Confederacy were twice as likely to die of a camp disease as from an injury incurred on the battlefield. Figure 1 demonstrates that these general conditions were specifically reflected in the South early on. Overall, of the estimated 258,000 Confederate deaths, only 94,000 were from combat. The popular image of hastily performed amputations and grizzly, bloody minie ball extractions, while not entirely without merit, needs considerable moderation in light of these statistics.

This is not the only misconception, however. The notion that Civil War medicines were completely ineffective is also inaccurate. While it is true that a failure to know or understand the microbial basis for infectious diseases seriously hampered diagnosis and treatment of many ailments, knowledge of curatives harvested, often locally, from natural sources benefited medicine during the war. For example, certain barks high in tannins, such a blackberry and logwood, were effective astringents against diarrhea. Willow and poplar barks contained salicylic acid (key constituent of aspirin), methyl salicylate (oil of Wintergreen) provided a useful liniment, opium (from Turkey or the Far East) gave palliative care, and, most important of all, cinchona bark (*Cinchona officinalis* of South America), the crude drug from which quinine sulfate was made, could prevent and relieve malaria.

Because the predominant medical adversary was disease, obtaining drugs became a major problem for the South. This was largely a result of the Union naval blockade. Lincoln's declaration of a blockade, an operation that Gen. Winfield Scott took and incorporated into a larger strategy designed to bring the South to its knees, would eventually have its intended effect. Although derisively dubbed "Scott's Anaconda" by the old warhorse's opponents, the term was more apt than many cared to admit, as the life was slowly squeezed out of the Confederacy's supply lines. Quinine sulfate, cinchona, and opium all became scarce imports in the South.

Surgeon General Moore attempted to solve this problem by implementing an alternative supply table utilizing indigenous plants for medicinal purposes. With the war not quite a year old, the surgeon general called for an effort "to diminish its [the South's] tribute to foreigners" by "the appropriation of our indigenous medicinal substances of the vegetable kingdom." Moore instructed the medical corps to comb their districts and collect and forward to the purveyors "the enumerated remedial agents, or others found valuable." This supply table of "indigenous remedies" was based upon the work of South Carolina physician Francis Peyre Porcher's *Resources of the Southern Fields and Forests* (1863), a work commissioned by Moore and distributed to medical personnel throughout the Confederacy. Porcher's *Resources* was—and remains—a monumental work. Listing some 35,000 native plants, Porcher believed that 410 had medicinal value. Moreover, he gave an exhaustive list of native substitutes for imported drugs.

In the end, however, few if any valuable substitutes were discovered. Most importantly, cinchona and its refined derivative, quinine sulfate, never found an equivalent in the South. The absence of opium too became seriously felt. As the war dragged on, these deficiencies appreciably affected the war effort, the impact of which cannot be overestimated. Writing in August 1864, the medical director of the Army of Northern Virginia, Lafayette Guild, felt (like many of his colleagues) the impact of medical store deficits and warned the surgeon general: "There are almost daily complaints from the medical officers particularly from those of General Beauregaurd's command of the great scarcity of medical supplies, especially such as tonics and antiperiodics, much disease at present prevails among the troops in the trenches consisting principally of Intermittent & Remittent Fever and bowel affection. Yet it is believed that the strength of the army would not be naturally influenced if suitable remedies could be obtained for the treatment of cases in camp. The scarcity of Sulphate of Quinea [i.e. quinine] is assuming a very serious character, all of its substitutes that can be procured, are being used yet with little permanent benefit. The necessity of sending so many patriots to General Hospital arises only from the want of medicines suitable for their treatment in camp & the field infirmaries."

Thus, while disease has often been described as a "third army," it increasingly became an ally of the North. With the Confederacy effectively strangled by the Union naval blockade, diseases such as malaria often went untreated. Furthermore, the sheer massive influx of new recruits from northern cities meant that, even with disease as a great leveler of troops in the field, the Confederates found their ranks sick and with few to no replacements. Time was not on the South's side. As the war dragged on, disease became an ever-increasing factor in debilitating the southern troops; the Union, in contrast, well stocked with supplies of antimalarials, could have regular "quinine call" to help prevent the disease among its ranks.

Despite massive challenges and deficits, medical advances were made by the Confederate Medical Department. Surgeon General Moore readily appreciated the one-story pavilion hospital, and its important innovations in improved ventilation and internal organization became an important advance in postwar hospital construction. Medical Director Samuel H. Stout developed a mobile hospital system for the Army of the Tennessee that in many ways was replicated in World War II. Similarly, the development of a decentralized supply system over the single-depot method proved effective and was carried into the future. The surgical lessons learned by surgeons and assistant surgeons on both sides of the conflict trained a whole generation of physicians in the value of practical bedside and surgical care over the former didactic training. Not just men laboring as physicians, but women too, like Kate Cumming and Sally Tomkins (who served as the only female commissioned officer in the Confederacy), would contribute to the emerging profession of nursing. It should be noted that a special report of the Confederate Congress in September 1862 indicated that hospital mortality was cut by half when patients were attended by female nurses.

In general, the war had a stimulating effect on training and education, medical inquiry, and practical health care. The clear benefits of a robust large-scale drug-manufacturing infrastructure over the impromptu efforts of the

South proved a catalyst for further growth and development and demonstrated the ineffectiveness of localized medical improvisation. In surgery new techniques were learned and old ones perfected, and for the first time general anesthesia could be applied on a mass scale, proving its safety and effectiveness. When famed Virginia surgeon Hunter Holmes McGuire claimed the administration of chloroform in 28,000 cases without a single death, physicians took notice and made the practice a standard of care south of the Mason-Dixon.

Perhaps H. H. Cunningham assessed medicine's wartime experience best when he wrote, "As a group Confederate medical officers labored valiantly for the cause in which they believed; it is clearly obvious in retrospect that their contribution to the military effort was indispensable." Despite a relentless "third army" strengthened by severe shortages, health-care providers in the South unquestionably brought back to their communities many soldiers who would otherwise have become tearful memories. Although germ theory of disease and major advances in therapeutics were still on the horizon, these apostles of Hippocrates could do no more than their times and trials allowed.

MICHAEL A. FLANNERY
University of Alabama at Birmingham

H. H. Cunningham, *Doctors in Gray* (1958); Michael A. Flannery, *Civil War Pharmacy: A History of Drugs, Drug Supply and Provision, and Therapeutics for the Union and Confederacy* (2004); Jeffrey S. Sartin, *Clinical Infectious Diseases* (April 1993); Glenna R. Schroeder-Lein, *Confederate Hospitals on the Move: Samuel H. Stout and the Army of the Tennessee* (1994), *The Encyclopedia of Civil War Medicine* (2008).

Climate and Weather

The American South is defined here as the region encompassing the states east of the Mississippi River and from Virginia southward. Both climate and weather play prominent roles on human behavior, including the cultural landscape and activities evident from clothing, architecture, and urban planning schemes. Climatic events occur more slowly from annual, decadal, to even longer time frames, commonly expressed as 30-year averages. Weather events occur hourly to daily, sometimes as extreme events such as severe flooding, snowstorms, and hurricanes.

Climatic characteristics are commonly expressed in terms of temperature and precipitation. Latitude is an important control on temperature, with warmer temperatures southward affected by higher values of net radiation. Also, seasonal temperature differences between summer and winter are greater inland because of increased continentality. For example, averaged January and

July temperatures at New Orleans, La., are 52.6 degrees and 82.7 degrees respectively. This is in contrast to a January average of 45.6 degrees and a July average of 88.7 degrees at Nashville, Tenn. Variations on temperature additionally result from topography, as the atmosphere generally cools with increasing altitude. The South's topography varies from sea level along the Atlantic Coastal Plain and Gulf of Mexico to more than 5,000 feet in parts of the Appalachian Mountains. Growing seasons are generally longer southward and along the coast. Beaufort, S.C., for example, has an average growing season of longer than 290 days, which enabled the establishment of the Sea Island cotton enterprises in the antebellum period.

Precipitation over the South varies from 40 to 60 inches annually. Mid-latitude cyclones in the winter and spring from the west, the Atlantic, and the Gulf of Mexico bring precipitation, occasionally in the form of snow. Summer precipitation results from surface heating of warm, humid air, which leads to thunderstorm activity. Topography affects the South's precipitation by physically forcing air masses to move orographically upward, thereby causing more rainfall on windward sides and less rainfall on leeward sides. For example, areas of the Blue Ridge Mountains occasionally experience heavy flooding because of the orographic uplift. Tropical storms and hurricanes affect the South mostly from June to October.

Dr. John Lining of Charleston, S.C., conducted the first systematic meteorological observations in the South that date back to 1738. Thereafter, into the early 19th century, weather observations were kept by medical doctors and surgeon generals of the U.S. Army because of their belief that climate influences health and disease, especially yellow fever epidemics. Some societies had weather observers, including the Medical Society of South Carolina in the late 1700s and early 1800s and the Charleston Board of Health and the Board of Health of New Orleans in the mid-1800s. Some detailed observations were kept by weather enthusiasts, such as those of the Smithsonian Institution; William Dunbar for Natchez, Miss.; Thomas Jefferson for Monticello, Va.; Nathan Smith for Washington, Ark.; Paul Tavel for Nashville, Tenn.; James Kershaw for Camden, S.C.; and Reverend Alexander Glennie for Pawley's Island, S.C. The U.S. Signal Service set up many weather stations in 1871, it being the forerunner of the U.S. Weather Bureau, which was established in 1891.

Plantation diaries and newspapers, dating back to the mid-18th century, also reveal daily weather information as experienced by southerners. Among the noted diarists are D. T. Merritt from Halifax County, Va.; Samuel Porcher Gaillard of Sumter District, S.C.; Daniel Cannon Webb from nearby Charleston, S.C.; William Nevitt near Natchez, Miss.; David Rees from St. Martin Parish,

La.; Valcour Aime of St. James Parish, La.; and Roswell King near South Hampton, Ga. Southern newspapers that contained weather information for many decades include the *Charleston Courier*, *Savannah Republican*, the *New Orleans Picayune*, the *Raleigh Register*, and *Arkansas Gazette*.

All of the South's long weather records reveal that climate has exhibited substantial variability with time. Prolonged warm-season droughts occurred in the Carolinas and Georgia during the early to mid-19th century in an unprecedented manner when just considering the record of the past 100 years. Longer growing seasons were evident at times in the early 19th century, enabling the growth of oranges northward into South Carolina. The winter of 1827–28 was extraordinarily warm throughout the entire South, with peach trees blossoming in many interior locations and ripe watermelons in North Carolina reported in January. This winter was extremely wet throughout much of the South, and areas along the Mississippi River experienced record flooding. A widespread killing frost followed in early April of that year, perhaps only surpassed by a similar widespread killing frost in April 1849 when considering records for the past 200 years. Other well-documented cold air outbreaks in the South occurred in February 1823 and February 1835. The winters of 1855–56 and 1856–57 were extraordinary cold throughout the Southeast, ranking within the top five coldest winters of the last few centuries.

Hurricanes have created a love-hate relationship to citizens of the South, and they also exhibited substantial variability in their activity through time. A levee and dike system to mitigate flooding hazard from hurricanes dates back to the antebellum period, and southerners often kept watchful looks for these "September Gales," sometimes trying to predict their occurrences through various scientific and nonscientific aspects. Very active periods of landfalling hurricanes in the South are evident in the 1830s, 1890s, and the mid-20th century. A distinctive inactive period of hurricanes is evident from around 1950–95, corresponding with vast coastal development and population growth. A return to a very active period is currently ongoing, as evidenced partly by four landfalling hurricanes in Florida in 2004 and the subsequent 2005 Atlantic hurricane season, which was the most active Atlantic hurricane season in recorded history and included Hurricane Katrina. Individual hurricanes have wrought heavy societal impacts in colonial and antebellum times for certain locations in the South. Some of these prominent events include major hurricanes that impacted New Orleans, La., in 1722, 1812, 1831, and 1856 and Charleston, S.C., in 1752, 1813, and 1893.

Global warming, resulting from anthropogenic emissions of carbon dioxide into the atmosphere, is expected to bring new challenges. Clearly, a longer growing season will result, but also likely are more extreme weather events and generally warmer temperatures. Society will have to adapt in terms of its agricultural, forestry, and water resource management, as well as anticipate increased drought and wildfire hazards.

CARY J. MOCK
University of South Carolina

Robert C. Aldredge, *Weather Observers in Charleston, South Carolina* (1940); James R. Fleming, *Meteorology in America, 1800–1870* (1990); David M. Ludlum, *Early American Winters II, 1821–1870* (1968); Cary J. Mock, in *Hurricanes and Typhoons: Past, Present, and Future*, ed. Richard J. Murnane and Kam-biu Liu (2004); Mart A. Stewart, in *Nature and Society in Historical Context*, ed. Mikuláš Teich, Roy Porter, and Bo Gustafsson (1992); Glenn T. Trewartha, *The Earth's Problem Climates* (1981).

Drug Use

The history of southern drug use falls into three distinct eras. Before the Civil War, southerners drank heavily but were largely untroubled by other intoxicating drugs, as was true of antebellum Americans generally. During the late 19th and early 20th centuries, while regional alcohol consumption declined, southerners used more opiates and cocaine than northerners and westerners. Then, during the mid- and late 20th century, the pattern reversed: southerners consumed fewer illicit drugs than other Americans, even as they confronted a range of novel drug-taking practices.

In the 19th century, southerners used opiates (including opium, laudanum, paregoric, and morphine) to relieve pain, alleviate anxiety, and check diarrhea. Widely prescribed by physicians, these drugs were also popular as home remedies. "Came home disheartened and miserable," Mary Chesnut confided in her journal in November 1861. "Was so ill I had to take morphine." Later she wrote, "After several weeks' illness—dawdling on, kept alive by Dr. T's opium—once more I was on my feet."

The trouble with such medication was that it could easily lead to addiction. In this, southerners were hardly unique: medical addicts could be found on both sides of the Mason-Dixon line. What set postbellum southerners apart was a rate of opiate addiction some 60 or 70 percent higher than the rest of the country. The difference reflected the prevalence of certain diseases, notably diarrhea, dysentery, and malaria. Being ill longer and more frequently than northerners, southerners resorted to opiates more often. Southerners also suffered proportionately higher Civil War casualties. Wounded and traumatized veterans, grieving parents and widows, and those afflicted with chronic diseases endemic to the region combined to form a relatively large pool of opiate-addiction candidates.

The situation changed during the early 20th century. As public health improved, as physicians became more conservative in prescribing, and as medical addicts gradually died off, a new group of users began to emerge: younger men who began using opiates to sober up after drinking sprees or as a source of dissipation. Sometimes called "pleasure users," they evoked less sympathy than the older, mostly female medical addicts.

Similar changes occurred in other parts of the country. What set the new breed of southern addicts apart was their continued use of morphine. In the 1920s and 1930s, when northern addicts had largely switched to black-market heroin, southern addicts were still injecting morphine. Heroin was less available in the South, and southern physicians were more willing to maintain addicts quietly. When a doctor "wrote scrip" for an addict he prescribed morphine, as heroin was illegal after 1924.

Most southern opiate addicts were white. Reports from Tennessee, Texas, and Florida showed blacks to be underrepresented among known users. To the extent that blacks lacked access to doctors, they were less likely to receive opiates, especially through hypodermic injection.

Southern blacks did use cocaine. Sometime in the late 1880s or 1890s, black stevedores in New Orleans began taking the drug as a pick-me-up while loading or unloading steamboats. The practice spread to cotton plantations, railroad work camps, and levee construction sites. Some employers stocked the drug and dispensed it in the manner of the old whiskey ration doled out to laborers. Cocaine also became a popular recreational drug among blacks, especially those who lived in or on the fringes of the urban underworld.

"Many of the horrible crimes committed in the Southern States by the colored people can be traced to the cocaine habit," Col. J. W. Watson of Georgia charged in 1903. Over the next decade, other white authorities made similar claims. The allegations may have been politically or racially motivated, as some scholars have suggested, but they could also have been based on actual events. Prolonged cocaine use produces erratic, paranoid, and sometimes violent behavior, as physicians then and now have noted. Given the supercharged racial atmosphere, it would have taken no more than a handful of genuine incidents involving blacks to create the specter of a "cocaine menace." Alternatively, when black cocaine users committed crimes for other reasons, such as obtaining money, police and journalists may have simply attributed their deeds to the cocaine.

Southerners also learned to associate marijuana with violence. Hemp, the

An illustration from 1900 depicting a drugstore with an elderly man, the pharmacist, dispensing a "Bracer" to a crowd of eager consumers, while a young girl secures a bottle of "Soothing Syrup." On the counter are bottles and packets of "Arsenic, Strychnine, Antipyrin, Nerve Stimulant, Opium, Cocaine" and "The Needle." Signs on the wall state "The Killem Quick Pharmacy," "Open all night," and "Prescriptions carefully compounded." The saloonkeeper leans against a column and laments that he cannot "begin to compete with" the drug trade. (Louis Dalrymple, artist, J. Ottmann Lith. Co., Library of Congress, [AP101.P7 1900], Washington, D.C.)

source of marijuana, had long been a cash crop in the South, notably in Kentucky, where farmers grew it for fiber, seed, oil, and medicine. During the 19th century, however, few southerners sought intoxication with the drug, which elicited no particular concern. The situation was otherwise in Mexico, where newspapers ran stories of prisoners and bored soldiers going wild under marijuana's influence. Mexico City banned its sale as early as 1869. The drug's unsavory reputation was established well before migrant Mexican workers introduced marijuana smoking into the American Southwest in the early 20th century.

By 1924, aroused legislatures in several states, including Texas, Arkansas, and Louisiana, had passed restrictive measures. Marijuana use had spread to the native-born population, including lower-class and criminal elements in Houston, Galveston, and New Orleans. Residents of these port cities encountered Caribbean sailors and immigrants, as well as itinerant Mexicans, who smoked the drug. New Orleans became a center for marijuana sales and distribution. A sensational 1926 newspaper campaign charged that marijuana smoking had become widespread in the city and that it had inspired dangerous behavior among youngsters of both races. Although Louisiana passed stricter laws in 1927 and again in 1935, reports of marijuana-inspired crimes persisted. Federal officials used these and other incidents to justify the 1937 Marijuana Tax Act, a measure designed to suppress trafficking.

Some southerners remained skeptical about marijuana's link to violent crime. Dr. J. D. Reichard debunked the marijuana-crime reports and recounted stories of black field hands in Kentucky who were formerly seen loading their pipes with dried flowering hemp tops and then lighting them up. "There was never the slightest suspicion that this procedure caused abnormal behavior," he noted. "This is particularly important since aggressive behavior by a colored person was, to put it mildly, viewed with alarm."

When Reichard wrote those words, in 1944, southerners worried little about drug abuse of any variety. Heroin and cocaine were everywhere scarce, due to wartime disruptions of supply. Many narcotic addicts had to quit or make do with alcohol, paregoric, barbiturates, marijuana, or Benzedrine. The illicit narcotic traffic revived in the late 1940s and early 1950s, and with it came a new type of addict: the young minority hipster infatuated with heroin as the coolest of drugs. Though many black heroin addicts were transplanted southerners—they or their parents had migrated to New York, Chicago, Los Angeles, Detroit, or another industrial city—few lived in the South itself. Even Baltimore, the largest southern city with 950,000 residents, had only a few dozen black hipster addicts at the close of the 1940s.

The number of heroin addicts in Baltimore and in nearby Washington, D.C., rose slowly during the 1950s and then more rapidly during the 1960s, when the nation experienced a surge in heroin addiction. By the end of the 1960s police and other investigators had identified about 4,000 Baltimore addicts. The younger ones pursued not old-fashioned junkie hustles and cons but brazen daylight muggings and robberies. White Baltimoreans, and black ones who could afford to join them, began exiting to the suburbs.

Even in the suburbs, southerners confronted an unprecedented increase in youthful drug use. In 1967 seniors at Milford Mill High School outside of Baltimore contented themselves with keg parties. In 1968 members of the school's self-proclaimed "Class of Grass" added pot parties to the senior social calendar. Smokers got younger over the next decade. In 1971, 7 percent of southerners aged 17 and under had tried marijuana. By 1979, 24 percent had. In 1977 alarmed parents in the Atlanta area founded Families in Action, the first of several "parents' groups" that lobbied against head shops, drugs in schools, and any policy or politician that tolerated drugs or minimized their danger to the young. Suburban, socially conservative, and Republican-leaning, the leaders of the parents' groups became influential allies of the Reagan administration when it renewed and hardened the drug war during the 1980s.

What happened in the South in the 1960s and 1970s was part of a national (in fact, international) increase in drug use tied to the postwar baby boom, the expanding counterculture, the popularization of novel drugs, and the growth of supply—which, in southern states, included drugs smuggled from Mexico and the Caribbean. Every national drug fad and trend—psychedelic experimentation; the abuse of amphetamines, barbiturates, and tranquilizers; the rediscovery of cocaine—eventually made itself felt in the region's cities and suburbs. Rural and small-town southerners showed less tolerance for drugs and their users, a prejudice dramatized when Louisiana rednecks harassed the long-haired bikers ("Yankee queers") in the hit 1969 movie *Easy Rider*.

If southerners could no more escape invading drug "bummers" than General Sherman's foraging variety of a century before, their overall rate of illicit drug use nevertheless remained lower than that of other regions. One in every four southern minors had tried marijuana in 1979, but that still compared favorably to more than one in three in the North and the West. Late 20th-century surveys consistently showed that, as a region, the South had the lowest rate of illicit drug use. Its rural, evangelical, and socially conservative character also translated into less per capita alcohol consumption and more resistance to liberal drug policies. As of 2011 no southern state had legalized medical marijuana.

Still, the South maintained its share of drug hot spots. New Orleans was one, Miami another—so much so that in 1990 the Drug Enforcement Administration designated Miami-Dade and surrounding counties as one of the nation's five high-intensity drug trafficking areas. Border states such as Oklahoma and Missouri developed thriving amphetamine subcultures, memorialized in Larry Clark's 1971 photo essay *Tulsa* and Daniel Woodrell's 2006 novel *Winter's Bone*, from which Debra Granik adapted the 2010 film. Pot farmers flourished in Tennessee and Kentucky, each of which produced more than $4 billion worth of marijuana in 2005. Florida pill mills diverted Oxycontin and other synthetic narcotics throughout the state and region. Users ground up the pills and then sniffed them or injected the dissolved powder for its heroinlike effects. "It's completely flipped," said the nurse-manager of a Jacksonville, Fla., methadone clinic, who watched her clientele shift from mostly heroin addicts in 1998 to mostly prescription pill abusers in 2010. More surprising still was the spread of pill diversion and addiction in Appalachia, a region formerly hostile to all illegal drugs save moonshine. It was as if the early 20th-century pattern of addiction, in which southerners used narcotics to blot out pain, anxiety, and chronic illness, had returned with a vengeance in the early 21st century. Only the face of southern addiction was no longer a sick old lady using opium or morphine, but a busted-up, chain-smoking coal miner who had become dependent on an even more powerful form of the drug.

DAVID T. COURTWRIGHT
University of North Florida

John C. Ball and Carl D. Chambers, eds., *The Epidemiology of Opiate Addiction in the United States* (1970); Richard J. Bonnie and Charles H. Whitebread II, *The Marihuana Conviction: A History of Marihuana Prohibition in the United States* (1974); David T. Courtwright, *Dark Paradise: A History of Opiate Addiction in America* (rev. ed., 2001); Lester Grinspoon, *Marihuana Reconsidered* (1977); Jill Jonnes, *Hep-Cats, Narcs, and Pipe Dreams: A History of America's Romance with Illegal Drugs* (1996); David F. Musto, *The American Disease: Origins of Narcotic Control* (1999); John A. O'Donnell, *Narcotic Addicts in Kentucky* (1969); Joseph F. Spillane, *Cocaine: From Medical Marvel to Modern Menace in the United States, 1884–1920* (2000); Charles E. Terry and Mildred Pellens, *The Opium Problem* (1928); Louis Vyhnanek, *Louisiana History* (Summer 1981).

Education, Medical

Southern medical education has always been shaped by the South's distinctive climate and disease environment, its medical profession and health-care institutions, and its educational system. All of these in turn have been strongly influenced by the South's racial and economic history of cash-crop agriculture dependent on slave labor, which set it apart from the urban Northeast, the power center of American academic medicine. Well into the 19th century, most aspiring southern physicians either apprenticed themselves to an experienced practitioner or attended the small, independent proprietary schools that proliferated beginning in the 1820s. Many chose to attend the better-established northern schools, and a handful studied in Europe. On the eve of the Civil War, most southern medical students at northern schools, including 300 in Philadelphia alone, withdrew to enroll in southern schools. By 1860 the South had established 23 medical schools, 15 of which were concentrated in Maryland, Virginia, the District of Columbia, Tennessee, and Kentucky. During the antebellum period, sectionalist physicians such as Samuel Cartwright embraced "states' rights medicine," which held that the southern environment had modified common diseases or created entirely new ones that could be understood and treated effectively only by physicians trained in southern schools. Southern medical professors and textbook authors were also among the chief proponents of biological black inferiority based on racist concepts of comparative anatomy and physiology, which would subsequently become widely accepted outside the South as well. Charity Hospital in New Orleans, the teaching hospital for the medical schools of Tulane University and later Louisiana State University, became a center of research on race-based differences in disease.

After the Civil War, both black and white physicians founded numerous proprietary medical schools. At this time, even Harvard Medical School had no admissions requirements, grades, or laboratory work, except for anatomy. The founding of the Johns Hopkins Medical School in 1893 as the first modern medical school, the formation of the American Medical Association Council on Medical Education in 1900, and Abraham Flexner's 1910 report to the Carnegie Foundation on medical education in the United States and Canada together ushered in sweeping reforms that made medicine the most expensive form of higher education and decimated the southern schools, particularly those for blacks. Between 1900 and 1923, the number of black schools shrank from 10 to 2, Howard University in Washington, D.C., and Meharry Medical College in Nashville, Tenn., which henceforth educated the vast majority of black physicians. Abraham Flexner concluded in his report on the South that medical education there was "trembling on the verge of extinction," but he still urged the General Education Board to fund specific measures to improve the schools, ranging from construction of laboratory facilities to establishing residency programs to creating a $500,000 endowment for teaching and equipment at Howard, which Flexner insisted "must not be permitted to disappear." Despite Flexner's low opinion of most southern schools, he and Carnegie Foundation president Henry S. Pritchett recognized the limitations posed by southern conditions and advocated tolerance of "greater unevenness." All the South's medical schools were handicapped by the serious underdevelopment of its public and higher education systems as well as its health-care institutions. Southern university endowments were uniformly low; the only three (Duke, Texas, and Vanderbilt) with more than $25 million before World War II also boasted the best medical schools. In 1914 Flexner, Pritchett, and the Southern Medical College Association threatened to remove southern schools from the American Medical Association Council on Medical Education's authority unless they were given time to adjust to the new admissions standards. Although Flexner, a native of Louisville, Ky., considered black physicians better suited to be hygienists than surgeons, he also made sustained efforts to secure major grants to improve southern medical schools for blacks as well as whites. By 1930 nearly all the southern schools required a liberal arts degree for admission and offered three to four years of graded instruction in medicine and surgery.

By World War II, of 76 medical schools in the United States, one-third were in the South, but most were at private universities, which severely limited access to medical education in the nation's poorest region. The dire shortage of health personnel, particularly those with even minimal public health training, was a major obstacle to delivering health services to the South's poor, widely scattered population. Physician-to-population ratios indicate the magnitude of disparity: in 1940 one physician practiced for every 592 persons in the Northeast and 755 persons in the nation, versus 1,064 persons in the South. For African Americans, the ratio was one black physician per 3,681 black population nationally, per 6,204 in the South, per 12,561 in South Carolina, and per 18,132 in Mississippi. Moreover, physicians were heavily concentrated in cities in a region that remained largely rural. Black physicians who were willing to give up their civil rights to practice in the segregated South were even less likely than white practitioners to locate in rural areas.

The Great Depression and World War II were catalysts for federal and state programs to address the major deficiencies in the South's health and educational systems. The South's high rates of maternal and infant mortality, low percentage of doctor-attended hospital births, and disproportionate number of World War II draft rejections all prompted southern medical societies and state legislatures to make increasing the supply of doctors and hospital beds a top priority. In Alabama and North Carolina, wartime health reform centered around campaigns to build a new four-year state medical school, as did postwar health reform in Florida and Kentucky. Progressive southern governors such as Earl Long of Louisiana and Sid McMath of Arkansas raised taxes to fund new public schools and enhancements to the state university system and also oversaw the refurbishing of state hospitals and mental institutions, all of which contributed to an improved environment for medical education.

As southern state governments grew more willing to spend tax dollars on health care and education, they also exhibited a hardening resolve to defend segregation. *Gaines v. Missouri* (1938) and subsequent Supreme Court decisions required public graduate and professional schools to admit qualified black applicants if no "separate but equal" school for blacks existed. Led by Florida Governor Millard F. Caldwell, the Southern Governors Conference created the Southern Regional Education Board (SREB) in 1948 to promote interstate cooperation and sharing of resources to improve and expand expensive graduate and professional programs for both black and white students. The primary focus of the SREB in its early years was medicine, dentistry, nursing, and pharmacy. To avoid building expensive new schools, southern states agreed to subsidize tuition for their residents at existing schools in other member states, including Emory, Tulane, and the University of Alabama. The largest recipient of SREB funding was Meharry Medical College, which received $2.5 million from 1949 to 1961 to aid approximately 500 black medical students and an additional $1 million for black dental students. Although the agreement with Meharry originated as a way for southern state governments to avoid desegregating medical schools, it made increasing the supply of black physicians a goal of southern public policy long before the advent of affirmative action or diversity programs, even after the 1954 *Brown* decision and white backlash. The SREB soon evolved to become a force for both racial equality and federal funding for medical education. In the late 1940s the National Medical Association and the NAACP mounted a joint campaign to end segregation in medical education, and Thurgood Marshall urged black students to apply to white medical schools. Between 1948 and 1951, the state universities of Arkansas, Texas, and North Carolina, the University of Louisville, and the Medical College of Virginia admitted their first black medical students, and by the late 1960s even schools in the Deep South were actively recruiting minority applicants with support from foundations such as the Commonwealth Fund and the Josiah H. Macy Foundation.

Major federal funding helped to more than triple the number of southern state medical schools from seven in 1940 to 23 by 1985, and existing schools also increased their class sizes. New southern schools accounted for much of the increase in the nation's supply of physicians during this period. The 1946 Hill-Burton Hospital Survey and Construction Act fostered the growth of institution-based, capital- and technology-intensive medicine in the South. Teaching hospitals represented about 20 percent of all Hill-Burton projects, including the burgeoning new academic medical centers at the state universities of Alabama, North Carolina, Texas, and Florida, as well as the expansion of facilities at private medical schools such as Duke and Bowman Gray (now Wake Forest). Funding from the Veterans Administration and the National Mental Health Act also underwrote construction on southern medical school campuses as well as support for residency programs. At the state level, addressing the shortage and maldistribution of physicians and improving health-care access for medically underserved rural and minority populations were the most common traditional justifications for founding new medical schools at public expense, but once established, schools like those at the University of North Carolina and the University of Florida had tended to abandon such priorities in favor of pursuing specialty training and clinical research. Rivalries between the larger, older flagship universities and newer institutions such as East Carolina University and University of South Florida contributed to political one-upmanship that sought to increase prestige and state budget lines via establishing medical schools. This tactic was facilitated by the populist fervor that overtook southern capitals following the Supreme Court's "one man, one vote" decision, which required southern states to comply with the Voting Rights Act of 1965 by redrawing racially gerrymandered voting districts.

In the context of the Great Society push for national health reform, the 1963 Health Professions Educational Assistance Act, Regional Medical Program (RMP) (1965), 1968 Health Manpower Act, and Area Health Education Centers (AHEC) residency program (1971) were particularly important in supporting the expansion of southern medical education as well as increasing the region's supply of primary care and minority physicians. AHEC, RMP, and other residency programs proved more successful than previous attempts to encourage M.D.s to consider general practice in rural and inner-city areas. These programs, along with the recruitment of numerous prominent faculty from outside the South to fledgling schools in Birmingham, Ala., or Gainesville, Fla., liberalized racial attitudes across campus and transformed southern medical schools from provincial outposts to modern health science centers. In 2000 Florida State University in Tallahassee established the first new allopathic medical school in the United States in more than 20 years. Accredited in 2005, the school prepares physicians for community-based practice and emphasizes primary care, geriatrics, cultural diversity, and the needs of underserved populations.

KAREN KRUSE THOMAS
Johns Hopkins University

James O. Breeden, *Bulletin of the New York Academy of Medicine* (March–April 1976); Walter E. Campbell, *Foundations for Excellence: 75 Years of Duke Medicine* (2006); John Duffy, *The Tulane University Medical Center: One Hundred and Fifty Years of Medical Education* (1984); Timothy Jacobson, *Making Medical Doctors: Science and Medicine at Vanderbilt since Flexner* (1988); Kenneth M. Ludmerer, *Time to Heal: American Medical Education from the Turn of the Century to the Era of Managed Care* (1999); William W. McLendon, Floyd W. Denny Jr., and William B. Blythe, *Bettering the Health of the People: W. Reece Berryhill, the UNC School of Medicine, and the North Carolina Good Health Movement* (2007); Tennant S. McWilliams, *New Lights in the Valley: The Emergence of UAB* (2007); James Summerville, *Educating Black Doctors: A History of Meharry Medical College* (1983); Karen Kruse Thomas, *Journal of African American History* (Summer 2003).

Environmental Health

Environmental health—the impact on human health and disease of both geophysical and built environments—covers a vast field, from disease ecology to climate change, agricultural policy to urban sprawl, pesticide resistance to pollution prevention. The World Health Organization offers one broad definition: "the direct pathological effects of chemicals, radiation and some biological agents, and the effects (often indirect) on health and well-being of the broad physical, psychological, social and cultural environment, which includes housing, urban development, land use and transport."

As elsewhere, from the moment of conquest, the interchange of pathogens inflicted disastrous consequences on indigenous societies. European colonizers found the hot, humid semitropical environment richly fertile but especially inhospitable, contributing to the very notion of the South as a separate region. Slave plantations dominated tobacco, cotton, rice, and sugar cultivation in the lower South, exacting a deadly toll on African slaves and depleting the landscapes their labor shaped.

As medical historian Margaret Humphreys has noted, the antebellum South was widely viewed as the nation's "most diseased region." The designation derived largely from the prevalence of infectious diseases, such as malaria, yellow fever, and cholera. As least through 1929, when historian Ulrich B. Phillips justified the South's embrace of racial slavery with a claim that the social system flowed logically from the Africans' resistance to malaria and purported natural ability to work in extreme heat, a crude environmental determinism largely reigned. Besides its explicit racial bias, the argument ignored climatic variation within regions of the South and overlooked other causal factors in disease: diet, housing and working conditions, the spread of pathogens, and human interventions in the landscape. The climatic rationale for claiming southern distinctiveness has long since been rejected, but, as historian Raymond Arsenault has noted, "Climate may not be the key to southern history, but climate does matter."

More multicausal explanations now prevail. The history of "natural" disaster in the South—the boll weevil's swath of destruction of the cotton crop, the massive erosion evident in the Dust Bowl's sweep across the southern plains—reveals that human attempts to manage nature can have disastrous consequences. Weather-related events from the Great Flood of 1927 to Hurricane Katrina in 2005 laid bare human failures: inadequate levee engineering, continuing racial and class segregation, and mismanagement.

Until the 1870s acute infectious disease was the primary cause of death among southerners. Thereafter, the South entered an epidemiologic transition that saw a rise in chronic, degenerative diseases of an aging population. In the post–Civil War period, as Humphreys also noted, yellow fever epidemics drove the development of public health boards and federal intervention in southern states, yielding advances in human health. As the mode of transmission of the malaria parasite became known in 1898, identification of the mosquito as the malaria disease vector facilitated a further reduction in disease. A fledgling public health and sanitation movement took hold in the South in the post–Civil War period, with reformers draining swamps, promoting sanitary sewers, and conducting educational campaigns. Nonetheless, the South continued to lead the nation in the prevalence of acute of infectious diseases, and the reputation retarded investment in southern industrial development.

Early 20th-century campaigns succeeded in reducing the incidence of hookworm and pellagra, also identified as scourges peculiar to the South. The impetus for improved health and sanitation was driven largely by employers' desires for a healthy work force. Health reformers often attributed the prevalence of disease to racial differences or the work ethic of the southern poor, failing to acknowledge the larger social determinants of disease. In its 1909–14 campaign to eradicate hookworm, a disease characterized by extreme lethargy, the Rockefeller Sanitary Commission sought to transform a population labeled "lazy southerners" to a work force groomed for industry. By 1926 pellagra researcher Dr. Joseph Goldberger linked the disease with a vitamin deficiency in the southern corn-based diet. Poverty, not temperament, blocked access to healthy foods and adequate nutrition.

Southerners' health was deeply affected by the environs in which they worked. The dominance of extractive cotton monocrop agriculture and the sharecropping system debilitated southern agricultural laborers, black and white. The well-being of textile workers, then mostly white, declined as they entered the dank, dust-laden cotton mills. Extractive industries like forestry and mining drew resources from the terrain and the people who worked the landscape. Deep pit miners suffered and died early from black lung disease. Some of the nation's worst cases of industrial disease occurred in the South. For example, in the early 1930s miners at Gauley Bridge, W.Va., digging a railroad tunnel through dense silica deposits, contracted the deadly lung disease silicosis. More than 750 mostly African American migrant workers lost their lives.

Industrialization reengineered the environment and led to greater urbanization, although the majority of southerners lived in rural areas until the 1960s. New technologies promised better health and enhanced quality of life for those who enjoyed access. Air-conditioning, invented by Florida physician John

Gorrie to improve the health of his patients, spread first as a technology for promoting workplace productivity. Big industrial projects, such as the Tennessee Valley Authority's system of dams to control flooding and generate electrical power, attempted to manage nature for regional development but also had consequences for human health. Rural electrification generally raised living standards and quality of life for those who could afford it.

Along the South's more than 3,000 miles of coastline, rich regional fisheries in the Chesapeake Bay and the Gulf of Mexico have suffered declines. Dead zones emerged in the 1970s in both estuaries. Owing to a condition known as hypoxia, caused by fertilizer and nutrient runoff from bordering farms, the seas lack sufficient oxygen to sustain aquatic life, threatening seafood safety and supply. Coastal waters were threatened as well by the 2010 BP (formerly British Petroleum) Deepwater Horizon oil disaster, which cost 11 lives and spewed an estimated nearly 5 million barrels of crude oil in the Gulf. Chemicals from the oil and the dispersants used to combat it have been measured in the bodies of coastal residents, but environmental health consequences remain largely unquantified.

By the late 1980s, the South became identified as "the nation's dumping ground." A 1987 United Church of Christ Commission for Racial Justice study, *Toxic Wastes and Race*, highlighted the disproportionate concentration of toxic waste sites in southern locales, also concentrated in communities of color. Though environmental injustice was not limited to the South, southern activists stood on the front edge of a wave of local protests against industrial pollution and toxic wastes that linked environmental conditions to human health. Advocates used mass protest, pressure on regulatory agencies, and toxic tort lawsuits to win redress. The strip of oil refineries and chemical plants between New Orleans and Baton Rouge, labeled Cancer Alley because of the elevated levels of different types of cancers found among the populations, became a key center of protests against new chemical manufacturing facilities.

These antitoxics battles highlighted activists' concerns over potential health effects of industrial pollution from heavy metals, such as lead and arsenic, and chlorinated hydrocarbons, including PCBs and DDT. In Triana, Ala., for example, DDT dumped into the Tennessee River from the Olin plant near Huntsville exposed the population to toxic amounts of the pesticide. In Warren County, N.C., civil rights groups and environmental activists joined forces to protest PCB dumping in a landfill near an African American neighborhood. Toxic caches of chemical weapons and hazardous waste sites located at military bases, a high proportion of which lie in the southern states, sparked an antimilitary toxics campaign.

In the context of a national legal system that provided measures to punish polluters after the fact—but with few precautionary laws to restrain chemical dumping before damage was done—the new wave of environmental activists connected health, environment, and civil rights movements that formerly stood apart. Key individuals included activist sociologist Robert Bullard, author of *Dumping in Dixie*, who mapped the disparate proximity of hazardous waste sites to the predominately African American communities in the South; Damu Smith of Greenpeace, who cofounded the Black Environmental Justice Network and helped steer the successful defeat of a large polyvinyl chloride plant near Convent, La.; and MacArthur Fellow Wilma Subra, a chemist who monitors chemical contamination of soils and waterways in African American, Latino, American Indian, and low-income communities.

Debate raged in law review articles and the popular press over whether race or economics drove the acknowledged disparities. Examining 10 studies on the siting of hazardous waste facilities, one scholar noted the "ambiguous and complicated entanglement of class, race, educational attainment, [and] occupational patterns," as well as rural-urban and market-siting dynamics, but found "race a better predictor . . . than income."

In the final decades of the 20th century, attention turned to the impact of the built environment on health. Sprawl and the lack of widespread access to quality low-cost food, housing, and medical care contribute to diabetes, obesity, and cardiovascular disease and tend to worsen health outcomes in general, but their effects are often indirect and difficult to document. An unintentional experiment in Atlanta highlighted the link between disease and urban transportation networks, when reduced highway air emissions during the 1996 Summer Olympics yielded lower asthma admissions at local hospitals. Environmental disparities are evident in the location not only of noxious facilities but also of beneficial amenities. A growing body of research shows the health benefits of living near green space and in walkable neighborhoods, to which communities of color and low-income residents disproportionately lack access.

Competing medical, scientific, and legal claims make it difficult to establish with precision the causal relationship between environment and disease. Scientific and medical certainty remain elusive, especially in regard to diseases with long latency and multiple, synergistic exposures. In such cases, assigning the legal burden of proof allocating responsibility for pollution control remains difficult. Translating ecological health findings to public health policy is made even more difficult in southern states, where there is a history of lax environmental enforcement.

The U.S. Census Region designated as the South maintains an unenviable position as the site of higher disease rates even today. The Centers for Disease Control and Prevention (CDC), based in Atlanta, is the nation's foremost entity for monitoring public and environmental health globally. In 2011, the CDC reported, the South led the nation in incidence of chronic diseases, including lung cancer, diabetes, stroke, heart disease, and obesity, all of which are influenced by environmental conditions. Recent developments in scholarship at the CDC and elsewhere suggest multicausal factors in environmental health. Researchers are exploring, for example, the effects of chemicals on hormones and endocrine disruption and the role of environmental factors in triggering gene expression related to disease, research that will have an impact on how future populations understand the American South, environment, and public health.

ELLEN GRIFFITH SPEARS
University of Alabama

Barbara L. Allen, *Uneasy Alchemy: Citizens and Experts in Louisiana's Chemical Corridor Disputes* (2003); Raymond Arsenault, *Journal of Southern History* (November 1984); Robert D. Bullard, *Dumping in Dixie: Race, Class, and Environmental Quality* (1990); Martin Cherniack, *The Hawk's Nest Incident: America's Worst Industrial Disaster* (1986); Pete Daniel, *Toxic Drift: Pesticides and Health in the Post–World War II South* (2005); Cheryl Elman and George C. Myers, *Demography* (November 1999); Margaret Humphreys, *Malaria: Poverty, Race, and Public Health in the United States* (2001), *Yellow Fever and the South* (1992); Barry L. Johnson, *Environmental Policy and Public Health* (2007); William A. Link, *Journal of Southern History* (November 1988); Todd Lee Savitt and James Harvey Young, *Disease and Distinctiveness in the American South* (1988); Mart A. Stewart, in *Nature and Society in Historical Context*, ed. Mikuláš Teich, Roy Porter, and Bo Gustafsson (1997); Allen Tullos, *Southern Exposure* (Summer 1978).

Eugenics

Eugenics is the set of scientific beliefs designed to improve the human race by selective breeding. Developed in the late 19th century, eugenics fused scientific methodology, racialized ideologies, and invasive state action to promote the goal of human betterment. American scientists, physicians, social workers, psychologists, educators, and politicians participated in this movement throughout much of the 20th century, creating a problematic legacy with far-reaching consequences. The social, legal, medical, and ethical questions raised by eugenics still have resonance today as we grapple with the implications of genetic testing and the human genome project. And many of the nationally contentious issues surrounding eugenics have been played out in the American South.

By the early decades of the 20th century, individuals began to apply the abstract science of eugenics to real world situations. Because eugenics advocates encouraged what they called "the breeding of the best," they had to develop ostensibly objective, verifiable criteria to identify those people. This aspect of eugenics, called positive eugenics, emphasized public exhibits and displays at places like the Georgia State Fair. Eugenic supporters organized Fitter Family competitions at such places, rewarding what they saw as superior families for having a "goodly heritage." However, the focus of eugenics was never on the encouragement of supposedly superior families to have more children. Instead, significantly more work and energy was placed into negative eugenics. This led to organized approaches to find, label, and separate those persons categorized as "feebleminded," those considered least worthwhile to bear and care for children. Using the resources of the Eugenics Record Office (ERO), located on Long Island and funded heavily by philanthropic foundations created by Andrew Carnegie and E. H. Harriman, eugenicists developed a two-pronged strategy to counteract what they saw as the menace of the feebleminded. Since feeblemindedness (along with other "undesirable" characteristics such as insanity, deafness, and epilepsy) were considered hereditary traits, eugenic supporters initiated a whole series of what they called pedigree studies, going back into the family histories of marginalized people to "prove" their inadequacies were hereditary. Once the fact of hereditary inadequacy was established, eugenicists then called for the prevention of future generations of these individuals, either by institutionalization in specialized facilities or by eugenic sterilization to "raise the intelligence of the state." Southern states provided a laboratory for both of these methods of eugenic control.

Many of the ERO pedigree studies examined poor rural southern mountain families and their relationship to crime, degeneracy, and feeblemindedness. A good example of this is Arthur Estabrook's *Mongrel Virginians*, published in 1926. The book exemplifies the close relationship between southern eugenics and concerns about racial integrity and miscegenation. Eugenic advocates used books like Estabrook's to drum up support for their programs of segregation and sterilization of those loosely labeled as feebleminded or defective. Between 1910 and 1923, nine southern states opened public institutions designed to house those populations. Established both to care for society's unfortunates and to control that same population as well, these institutions quickly became dumping grounds for petty criminals, sex offenders, and miscellaneous deviants. Hampered by traditional southern concerns about government spending, the facilities never served more than a fraction of those individuals considered dangerous to southern society. Institutional leaders and concerned government officials and citizens therefore searched for other, more efficient methods of controlling the perceived hereditary menace of the feebleminded.

By the mid-1920s Virginia leaders, in conjunction with the ERO, formulated a strategy to incorporate eugenic sterilization into their struggle against hereditary feeblemindedness. Indiana had legalized such measures as early as 1907 and other states followed suit, but few people were sterilized under these statutes, and courts had not ruled on their constitutionality. After a contentious 1917 lawsuit, in which a female patient sued the Lynchburg Colony (Virginia's state facility housing the "feebleminded") for damages resulting from her sterilization at the institution, Virginia legislators passed a model sterilization bill drafted by ERO director Harry Laughlin with the explicit goal of testing its constitutionality. In 1924 Lynchburg administrators picked Carrie Buck, a 17-year-old white resident of the colony, as the first person to be sterilized under the new statute. Buck's mother, Emma, had also been a resident of the Lynchburg Colony, and Carrie was institutionalized there after giving birth to an illegitimate daughter. To Virginia officials, the family exhibited the hereditary nature of feeblemindedness that could be stopped only by invasive measures. After a three-year court battle, the case reached the United States Supreme Court, which ruled 8–1 that Virginia had the right to compulsorily sterilize individuals who pose a hereditary threat to the state. The case, known as *Buck v. Bell*, was decided on 2 May 1927 with the decision written by Associate Justice Oliver Wendell Holmes. Holmes's short and rather simplistic statement contained the infamous rationale that "three generations of imbeciles is enough." Within five months, Lynchburg Colony doctors sterilized Carrie Buck. The case opened the door for eugenic sterilization to become legalized throughout both the region and the nation. But the procedure never became the panacea eugenic advocates hoped it would be. While some southern states readily followed Virginia's lead and passed similar laws, others, including Florida, Louisiana, and Tennessee, never enacted such legislation. Religious beliefs and concerns about invasive state action were major reasons that these states did not enact sterilization legislation. Figures regarding the numbers of eugenic sterilizations actually performed in the South are notoriously inexact. Best estimates conclude that southern states sterilized approximately 20,000 individuals for eugenic reasons from the 1920s until the 1970s.

The science of eugenics and its applications changed dramatically after World War II. In 1942 the United States Supreme Court ruled unanimously in

Skinner v. Oklahoma that Oklahoma's law allowing for the eugenic sterilization of certain types of criminals was unconstitutional. Though this decision did not directly relate to the sterilization statutes passed by southern states (which dealt with the sterilization of those labeled as mentally ill or "feebleminded"), it certainly changed the social framework in which eugenics operated. Combined with both the revelations of Nazi atrocities perpetrated in the name of improving the race by removing those considered inferior and more sophisticated genetic studies that showed "feeblemindedness" was not the simple discrete trait defined by the ERO, the *Skinner* decision marked a turning point in the history of eugenics. However, it is simplistic and naïve to suggest that eugenics simply disappeared after World War II. In fact, in the South, particularly in Georgia and North Carolina, authorities eugenically sterilized more individuals after the war than before it.

Moving away from strictly scientific reasons for sterilization, eugenic supporters emphasized economics and race as they developed new rationales for sterilizing those individuals viewed as unfit to reproduce. Clarence Gamble, heir to the Procter and Gamble fortune and a graduate of Harvard Medical School, provided funding and support for the continuation of the North Carolina program in the postwar years. In 1947 he published a poem entitled "The Lucky Morons" that exemplified his position. The poem concluded, "and because they had been sterilized, the taxpayers of North Carolina had saved thousands of dollars and the North Carolina morons lived happily ever after." Gamble's program, however, reached far beyond the walls of the institution. Under the auspices of the North Carolina State Board of Health, officials increasingly targeted poor black women still living in their own communities for sterilization procedures. By the mid-1960s, thousands of them had been sterilized, often tied to concerns over increasing benefit costs and burgeoning welfare rolls. Though about 40 percent of those sterilized in North Carolina were black, that number jumps to more than 65 percent when considering just those sterilized in the 1960s and 1970s. The gendered dimension is even more striking—85 percent of those sterilized in North Carolina were female. Complicating this issue was the simultaneous attempt to introduce birth-control methods to poor southern women of both races. A strange political coalition arose to oppose sterilization and birth control—with conservatives concerned about government-authorized birth control and liberals angered by what they viewed as a eugenic assault on the poor, especially African Americans. Not until 1974 did these programs finally officially end in North Carolina, but the issues raised by them still remain. That same year, in *Relf v. Weinberger*, a case brought on behalf of two black Alabama teenagers who had been sterilized by the state of Alabama without their consent, a U.S. District Court ruled that federal funds could not be used to fund eugenic sterilizations.

In 1972, though the Supreme Court had not overturned its *Buck v. Bell* decision (and to this date still has not), the state of Virginia repealed the 1924 sterilization statute upon which the decision was based. Five years later the state removed all mention of eugenic sterilization from its state codes. At the turn of the 21st century, eugenics again came into the public eye as victims of eugenic sterilization and their advocates pushed for public apologies and even compensation from government officials. In 2002 the Virginia State Assembly passed a resolution condemning eugenics and Virginia's participation in it. A year later Gov. Mark Warner remarked upon the "Commonwealth's sincere apology for Virginia's participation in eugenics," concluding that "the eugenics movement was a shameful effort in which state government never should have been involved." In 2003 South Carolina followed suit with a public apology, as did Georgia in 2007. A contentious debate has taken place in North Carolina, where in 2009 a historical marker was placed in Raleigh to mark the state's participation in eugenics. In 2011 the state held hearings before a Eugenics Compensation Task Force to determine if surviving victims of sterilization were deserving of financial payments in consideration of the damages caused by state action. The hearings centered on the 5,364 persons sterilized between 1946 and 1974. As of July 2011, the state legislature is still deciding whether this compensation is warranted. Therefore, although eugenics no longer plays a prominent role in state action, its dark shadow still covers many places in the South.

STEVEN NOLL
University of Florida

Susan Burch and Hannah Joyner, *Unspeakable: The Story of Junius Wilson* (2007); Gregory Dorr, *Segregation's Science: Eugenics and Society in Virginia* (2008); Pippa Holloway, *Sexuality, Politics, and Social Control in Virginia, 1920–1945* (2006); Edward Larson, *Sex, Race, and Science: Eugenics in the Deep South* (1995); Paul Lombardo, ed., *A Century of Eugenics in America: From the Indiana Experiment to the Human Genome Era* (2011), *Three Generations, No Imbeciles: Eugenics, the Supreme Court, and Buck v. Bell* (2008); Steven Noll, *Feeble-Minded in Our Midst: Institutions for the Mentally Retarded in the South, 1900–1940* (1995); Johanna Schoen, *Choice and Coercion: Birth Control, Sterilization, and Abortion in Public Health and Welfare* (2005); J. David Smith, *The Eugenic Assault on America: Scenes in Red, White, and Black* (1993).

Folk Medicine

Like the phrases "folk music" and "folk religion," "folk medicine" is a slippery term, suggesting shared practices of individuals excluded from fully evolved forms of cultural knowledge and expression. But if we interpret this term to mean any community's informal set of interrelated patterns of understanding and practice concerning wellness, folk medicine becomes a more universal and potentially sophisticated concept. Whereas formal medical practice derives its authority institutionally in the form of official regulation and credentialing, folk medicine relies on an authority that is typically relational in origin: the familiar practices and beliefs of a group both in the present and extending back through time. This relational authority is especially persuasive when it takes the form of testimony from trusted individuals in the family or community, who not only administer traditional treatments but also bear witness to their efficacy based on their own experience. Home remedies and many forms of informal treatment persist, supported both by their intrinsic medical value (which may vary) and by their connection with the life events and values of family and community through time. Understood in this sense, most of us are "folk medical practitioners" under some circumstances, whether relying on a spoonful of sugar as a hiccup remedy or settling a cough with a homemade toddy.

General attitudes toward *southern* folk medicine, again like those toward southern music and religion, have tended to both essentialize and marginalize the category, reflecting long-standing tensions between mainstream America and southern culture. To the extent that the term "folk" encodes "ignorant," "isolated," and "quaint," folk medicine has been inextricably bound up with popular stereotypes of the region. Hence many practices such as herbalism, planting by the moon, and others common in many parts of the United States are regarded as characteristically southern. Nevertheless, the oft-noted reverence accorded custom and communal-familial relationships in southern culture supports the observation that long-established traditional forms of treatment receive special attention and even privilege. The apparent conservatism of vernacular medicine in the South may be as much a matter of choice as necessity, giving older modalities such as herbal practice an enduring attraction that survives intact in the region, even as their popularity waxes and wanes in other parts of the United States.

What, then, are some of the distinctive characteristics of southern folk medicine? Ethnographers have certainly long observed a common preference for so-called natural (i.e., herbal or botanical) modalities. The range of plant life of the region is exceptionally diverse, lending itself to extensive botanical patterns of medical treatment in American Indian practice as well as adaptation by European settlers and enslaved Africans. The precise dynamics of communication of knowledge between these groups has been difficult to document, differing from region to region and confused by complex social factors involving race and power. Nonetheless, it is clear that such communication did occur, surviving, for example, in the still-prevalent preparation of sassafras (*Sassafras officinalis*) as an early spring tonic among southerners of different backgrounds, though it was originally an Indian practice. Plants in common medicinal use among Indian, European, and black southerners have included wild garlic (*Allium reticulatum*), May apple (*Podophyllum peltatum*), mullein (*Verbascum thapsis*), blackberry leaves (*Rubus villosus*), cotton root bark (*Gossypium herbaceum*), American senna (*Cassia marilandica*), wild yam (*Dioscurea villosa*), and Seneca snakeroot (*Polygala senega*), among many others. Although the healing properties of herbs are understood by healers of all backgrounds to inhere in the plant itself, there is also a widespread belief that these properties and discernment in their use are both God-given gifts, intertwining natural and spiritual folk medical belief.

The conservative preference for older vernacular forms of treatment extends to over-the-counter products such as Syrup of Black Draught and 666 Laxative Tonic—medicines that are not necessarily manufactured in the South but still enjoy a ready market there. Preference for older forms of administration also remains: pungent petrolatum-based topical ointments such as Cloverine (1860) and Porter's Liniment Salve (1871) and analgesics such as BC Powder (1906) retain their popularity in many small-town southern pharmacies. While not "natural" in the strict sense, these over-the-counter remedies pointedly advertise their botanical content. Their packaging frequently employs typography and graphics that are self-consciously "old-time," depicting or suggesting transgenerational family use.

Just as southern folk medical practice includes commercial products, undermining the stereotype of exclusively home-prepared "dooryard" remedies, long-standing patterns of written transmission subvert popular assumptions concerning southern folk medicine as exclusively oral. From a very early period, an informal written tradition has been maintained: early plantation owners and other householders often kept "commonplace books," in which illnesses and effective remedies were carefully described. An extension of this practice is evident in the lively tradition of privately published compendia of remedies and herbal advice that are continued to this day, exemplified by such works as Tillman Waggoner's *Poor Man's Medicine Bag* (1984).

Commercially published works also served to support and reinforce botanical treatment. Southern practitioners in even the remotest regions often made use of books and almanacs in making medical choices. Among the most influential of such works was Dr. John C. Gunn's *Domestic Medicine, or Poor Man's Friend* (1830), a compilation of medical advice that included an extensive section on healing plants that could be discovered or cultivated on the Kentucky-Tennessee frontier. The popularity of this work extended well beyond the Mid-South. At one time, it was said that no southern household was complete without a Bible, an almanac, and a copy of Gunn's work; late editions (the last published in 1920) can still be found in some homes, with careful handwritten marginal notations. More recently, relevant volumes of Eliot Wigginton's *Foxfire* compilations of folklife and oral history are sometimes used for domestic reference, along with other well-known published herbals.

In addition to "natural" remedies, attention to astrological phases of the moon has been, and remains, a factor not only in planting and harvesting but also in making human and veterinary health-related decisions, from the timing of tooth extraction to the gelding of livestock. Along with gardening tips, best fishing dates, tide schedules, and weather prognostication, moon phase information is available in regional and national almanacs, some of which are among the country's oldest serial publications. Farm supply businesses as well as funeral homes customarily offer their patrons complimentary wall calendars providing the same data. While not universal, the system of lunar beliefs underlying these publications crosses social, economic, and educational lines to a striking degree. The quasi-scientific basis for calendar-based customs is generally attributed to a kind of lunar magnetism. The concept of health as dependent on an unimpeded flow of magnetic energy cycling through the body represents a related health belief model, resulting in the popularity of homemade and manufactured copper bracelets as a preventive or cure for joint pain.

As a supplement to domestic folk medical practice, community-recognized traditional healing specialists represent a distinctive ongoing aspect of southern folk medicine. Individuals who carry on the traditions of "blowing out fire" (healing burns), "stopping blood" (halting hemorrhage), and curing warts or thrush (oropharyngeal candidiasis) generally have had the "gift" passed on to them according to established rules and utilize a combination of magical verbal spells and other techniques to bring about their results.

Straddling the categories of "natural" and "magico-religious" custom, the continuity of West-African-derived practices in the South have long attracted the attention of scholars, both among southern African Americans and in black urban populations of southern origin. Among the most dramatic of these are spells and practices relating to hoodoo and conjure medicine, often the territory of "root doctors" and other specialists. Although the practice is generally associated with African American tradition, there is no doubt that both white and black southerners have had recourse to such practitioners. Specifically African American health belief systems include concepts of internal balancing of blood factors including pressure, "sweetness," and other elements. The lexicon for these beliefs is complex and often a source of confusion in formal medical contexts.

Finally, no account of the spiritual aspect of southern folk medical practice would be complete without reference to the importance accorded to the healing power of prayer. This emphasis placed on the possibility of divine intervention in medical settings is evident in private domestic settings and hospitals, as well in as more public settings such as church services, gospel music programs, and healing revivals.

The stereotyping of southern folk medicine is as worthy of attention as the practices themselves. This process of stereotyping has not been static but rather has been shaped and reshaped over time by shifts in attitude toward the region. The emergence of allopathic medicine in the late 19th and early 20th centuries as the institutionally privileged form of treatment coincided with the conceptual evolution of the South as a discrete regional entity in American consciousness, as well as with a general postbellum antagonism toward the former rebels. Fictional and nonfiction accounts of southern folklife of this period include an alarming array of degenerate preacher-healers and demented granny midwives, practicing far from, or in defiance of, more enlightened forms of treatment.

Toward the mid-20th century, more extensive interaction of mainstream culture with the South only alleviated this demonization to the extent that practitioners became figures of fun. Female domestic healers were especially stigmatized by caricature, resulting in representations in comic strips such as Snuffy Smith's bumbling wife, Loweezy (from the pen of Billy DeBeck), and Li'l Abner's redoubtable mother, Pansy Yokum (from Al Capp). Actress Irene Ryan's portrayal of Granny Clampett, purveyor of "spring tonic," on the 1960s television show *The Beverly Hillbillies* continued the ignominious tradition.

By the 1970s, however, widespread social malaise attributed to the overindustrialization and overinstitutionalization of American society in general and formal medicine in particular brought about a reevaluation of southern folk medicine in the popular mind. As represented in popular nonfiction works such as Eliot Wigginton's *Foxfire* series, "matters of plain living" became matters of widespread interest, not least in the area of health. Real-life individuals such as *Foxfire*'s Aunt Arie Carpenter and the extensively studied North Carolina herbalist Tommy Bass, as well as the African American midwife Onnie Lee

Logan, have elicited respect and attention for the contrast their lives and practices offer to current health-care norms: an implicit critique centering on their emphasis on simple "natural" cures as well as their attention to relational issues.

As the classic distinctions between southern culture and the "American way of life" enumerated by C. Vann Woodward in 1960 continue to blur, the future of folk medicine as an aspect of southern folklife remains to be seen. New patterns of immigration from South and Central America and Asia, increased urbanization, and changes in economic and social patterns will surely result in adaptation of some traditions and elimination of others. But the careful observer may note that some shifts are more apparent than real: the pungent ointment Tiger Balm—invented by a Burmese herbalist in the 1870s and currently manufactured in Singapore—now comfortably takes its place on the shelf next to Cloverine and Porter's Liniment Salve, and acupuncture and other foreign energy-based modalities are readily understood and accepted through their similarity to the accustomed health belief models underlying chiropractic adjustment and the tried-and-true therapeutic copper bracelet.

ERIKA BRADY
Western Kentucky University

Erika Brady, *Healing Logics: Culture and Medicine in Modern Health Belief Systems* (2001); Anthony Cavender, *Folk Medicine in Southern Appalachia* (2003); John K. Crellin and Jane Philpott, *Herbal Medicine Past and Present: Trying to Give Ease* (1997); Sharla Fett, *Working Cures: Healing, Health, and Power on Southern Slave Plantations* (2007); John C. Gunn, *Domestic Medicine, or Poor Man's Friend* (1830, 1986); James K. Kirkland, Charles Sullivan III, Holly Mathews, and Karen Baldwin, *Herbal and Magical Medicine: Traditional Healing Today* (1992); Kay K. Moss, *Southern Folk Medicine, 1750–1820* (1999); Loudell Snow, *Walkin' over Medicine: Traditional Health Practices in African-American Life* (1993); Virgil J. Vogel, *American Indian Medicine* (1970); C. Vann Woodward, *The Burden of Southern History* (1960).

Gender and Health

Gender is a crucial factor in understanding the history of health status and health care in the American South. Gender studies examine the social construction of womanhood and manhood to explore how differences have been created and given meaning. This essay illuminates the gendered, and racialized, nature of health and healing. Through a focus on several case studies, it demonstrates the significance of reproduction in the history of women's health, war in the history of men's health, and public health in the history of women's health work.

In 1809 Jane Crawford of Kentucky rode 60 miles on horseback to be operated on by Kentucky physician Dr. Ephraim McDowell. Jane, a 47-year-old white woman, suffered from a rapidly growing ovarian tumor. At the time there was no reliable treatment so she agreed to an ovariotomy, a risky type of abdominal surgery. Dr. McDowell operated on Jane to remove her ovaries and found a cyst or tumor, which turned out to weigh an astonishing 22 pounds.

Responses to the gendered nature of this surgery were mixed. Jane must have been pleased, for she apparently made a complete recovery. However, some physicians were appalled that the surgery had taken place. They thought that gynecological surgery in general, and this procedure in particular, was too dangerous and too painful for white women to undergo. They believed that women's physical frailty meant that they could not tolerate such pain. Furthermore, they thought that such procedures violated female modesty.

Southern physicians' views of women's bodies were shaped by racialized views of womanhood, as indicated by the history of surgical experimentation. Not all women were included in gendered social conventions regarding female modesty and frailty. For example, Dr. J. Marion Sims, one of America's foremost surgeons who helped to develop the field of gynecology, launched his successful career through experimentation on the bodies of enslaved women. Although scholars know a great deal about Dr. Sims, they know almost nothing about these slave women whose role was absolutely essential to his development of new surgical procedures.

Sims practiced medicine in South Carolina and Alabama with little success until he moved into the field of surgery and developed the first reliable technique to repair vesicovaginal fistulas. Some women faced this dreaded condition because of lacerations following long or difficult labor or the misuse of forceps in childbirth. Tears then developed in the perineal tissues or in the walls of the vagina, bladder, or even rectum. Such fistulas permitted urine or sometime feces to constantly leak through the vaginal opening. It was very distressing for the women who had to live with this condition. They faced severe skin irritations, emitted unpleasant odors, and usually found themselves ostracized by others.

Slavery and the racialized views of black women's bodies made possible Sims's experiments. Like most white physicians, he was not concerned about issues of sexual modesty and the pain tolerance of African American women when he performed a series of experiments from 1845 to 1849 on several enslaved women who suffered from fistulas. Evidently in one case he even had to purchase his patient in order to operate on her. At a time when anesthesia was only just starting to gain acceptance for pain relief, he conducted extremely painful experiments repeatedly on black women without anesthesia. Three slave women in particular, Anarchia, Betsy, and Lucy, served as his subjects. Although they may have hoped his efforts would provide a cure for their health problem, they endured years as his surgical subjects while he invited other doctors to witness his operations. At times a dozen men stood around watching as Sims operated on the women, who were placed in humiliating positions on their hands and knees on a table in his front yard in Montgomery, Ala. He even trained the subjects of his experiments to assist him when other doctors lost interest. After at least 40 attempts, he finally succeeded in repairing the fistula by using pewter spoons to make an early type of speculum and using wire sutures to sew up the holes. Apparently a few white women came to Sims for treatment of their vesicovaginal fistulas after his success, but none of them were able to endure a single operation. It was simply too painful, raising more questions about what the development of this procedure must have been like for the enslaved African American women.

Even as 19th-century doctors debated the appropriateness of surgery to address women's reproductive health problems, the American Civil War presented surgical dilemmas regarding the best way to address men's health needs as soldiers. Wars have had profound effects on the development of gendered health professions and health care.

The Civil War led to an unprecedented level of private and government involvement in medical issues. In the South, several thousand doctors served with the Confederate forces, and the Confederate government built up a military medical establishment to cope with wartime health problems. Doctors provided some medical inspections of troops and tried to address camp sanitation issues; however, they faced difficulties obtaining medical supplies. Civilians organized medical relief organizations to furnish volunteer nurses. The government drew on nurses, doctors, and pharmacists to organize ambulance wagon teams, hospital trains, and hospitals to care for soldiers.

As battles raged across the South, improvements in weaponry made amputation the most frequently performed military medical operation during the war. Amputation became the standard treatment for fractures or broken bones and severely injured limbs. Indeed, the term "sawbones" to describe a surgeon dates from the war. However, amputations were not uniformly embraced as the appropriate solution to the needs of wounded men. Some physicians were appalled at the actions of overzealous doctors, many of whom gained their first surgical experiences on the bodies of young men during the war. Critics viewed these surgeons as simply butchers who mutilated bodies, hacking off limbs. They argued that doctors had other alternatives but could not be bothered to choose them.

Meanwhile, other physicians argued that amputations were the appropriate treatment for severe appendage wounds in order to save lives in the context of a war. At this time, doctors were unable to treat the men once infection developed, so they had to act fast, and an immediate amputation gave the patient the greatest chance for survival. Doctors performed amputations immediately to lessen the men's suffering. Men who were still in shock from their wounds and who were under the influence of alcohol or opium pills did not groan and scream nearly as much as men whose wounds were probed two to three days later. In addition, doctors chose amputation as the treatment of choice because of the sheer number of cases they had to treat. Battlefields produced simply too many wounded men at once, and it was easier to saw off an arm or leg than to do delicate surgery in the midst of a war. Indeed, excess time with one patient might mean the death of another. Finally, in a wartime situation doctors and patients might be required to relocate at a moment's notice so the doctor's task was to act quickly to save as many lives as possible. Hence, medical decisions and military pressures contributed to the high rates of amputations on the bodies of men.

The devastation caused by the Civil War affected men's health for years to come. Indeed, soldiers experienced widespread postwar mental and physical health problems. Furthermore, the huge number of amputations during the war stimulated the refinement and manufacture of prosthetics or false limbs for the now "disabled" veterans.

Gender shaped not only the history of treatments to repair women's and men's bodies but also the body politic through public health initiatives. Men acted to improve the health of communities and reduce preventable deaths as leaders of professional organizations and paid government health officials, while women's greatest contributions emerged in their volunteer health work. Two examples illustrate the significance of women's health work in the South: white women's reform efforts in Appalachia, and black women's health activism in Mississippi.

From the late 19th to the early 20th century, white women's health activism was central to efforts to improve the health and well-being of some of the South's poorest populations. Women promoted scientific medicine, the skills of public health nurses, and modern values in communities across Appalachia, including Kentucky, West Virginia, and Virginia. White, middle-class clubwomen organized efforts to provide modern health care to mining camps, while elite settlement workers provided health services to isolated mountain communities. As part of their health work, women volunteers paved the way for the displacement of traditional healers and midwives and greater acceptance of the authority of physicians. During the 1920s in Appalachia, however, doctors in their medical associations and doctors' wives in female auxiliaries of state and county medical associations tried to shut women volunteers out of community health work by dismissing their claims to community health knowledge.

In a parallel fashion, black women's clubs launched community health work, including clean-up campaigns, as part of a black health movement from 1890 to 1950 across the South. In 1915 Booker T. Washington nationalized such efforts in launching an annual observance of National Negro Health Week, supported by African American teachers, ministers, clubwomen, businessmen, midwives, nurses, dentists, and physicians. Female lay workers continued to play an important role in black health work even after the health-care professionalization of the 1920s because segregation and racism severely limited the number of black health professionals.

One of the most impressive volunteer health projects for African Americans was the Alpha Kappa Alpha Mississippi Health Project from 1935 to 1942. The project was designed, financed, and carried out by Alpha Kappa Alpha, the oldest black sorority. For several weeks during the Great Depression Dr. Dorothy Boulding Ferebee led a dozen middle-class sorority volunteers in a health campaign for poor black sharecroppers in the Mississippi Delta. Each summer 3,000 to 4,000 people attended the clinics, which provided medical examinations, vaccinations, and health education. The effect on the health of children was perhaps most significant. The volunteers provided well over 15,000 children with immunizations against such devastating diseases as smallpox and diphtheria.

SUSAN L. SMITH
University of Alberta, Canada

Sandra Lee Barney, *Authorized to Heal: Gender, Class, and the Transformation of Medicine in Appalachia, 1880–1930* (2000); Estelle Brodman and Elizabeth B. Carrick, *Bulletin of the History of Medicine* (Spring 1990); Eric Dean, *Shook Over Hell: Post-Traumatic Stress, Vietnam, and the Civil War* (1997); James H. Cassedy, *Medicine in America: A Short History* (1991); Kenneth M. Ludmerer, *Learning to Heal: The Development of American Medical Education* (1985, 1996); Deborah Kuhn McGregor, *Sexual Surgery and the Origins of Gynecology: J. Marion Sims, His Hospital, and His Patients* (1990); Judith M. Roy, in *Women, Health, and Medicine in America: A Historical Handbook*, ed. Rima D. Apple (1990); Susan L. Smith, *Sick and Tired of Being Sick and Tired: Black Women's Health Activism in America, 1890–1950* (1995).

Healers, Women

Women in the South have a long tradition of helping family and friends maintain and restore health. Healing traditions were brought to the South with the early settlers, and they evolved as ideas and procedures were incorporated from European medical practice, African traditions, and American Indian traditions. The passing of remedies and techniques for care of the sick through generations of women is found in many cultures. Distinctive southern healing characteristics stem from the types of rural areas in which folk medicine practices have predominated and from the healers' use of indigenous plants and animals. Because modern techniques for controlling infectious diseases were adopted later in the South than in the North and because many rural southern areas have had shortages of medical personnel, folk healing practices have been particularly important throughout the region.

A mainstay of southern healing has been the use of readily available ingredients. Traditionally, many rural women raised herbs and medicinal plants such as comfrey, ginger, and catnip, along with their flowers and vegetables. These movements also made use of wild plants and trees, such as cottonwood leaves and fever grass. Household staples—eggs, baking soda, sugar, and whiskey—were also important to the care of the infirm. Teas and syrup were the vehicle for many medicines; other agents were directly applied as poultices to sores, sprains, and pains.

The role of women in healing and nursing was strongly supported by tradition. In years when infectious diseases such as smallpox, typhoid, and malaria caused much illness and death, women spent long hours alleviating sickness and suffering. Death was a commonplace occurrence, but religious beliefs and the belief in a joyous afterlife helped to ease the pain associated with death. Before the 20th century, medicine was not a well-established profession based on scientific principles. Until the reform of medical education around 1910, almost anyone could claim competence and practice medicine. Mistrust of and disdain for doctors were widespread, as were social movements to restore the art of healing to the domestic sphere. Health-reform movements flourished throughout the country in the mid-19th century because of the failure of medical professionals to cure and because women sought a way to improve the quality of life in a confusing world undergoing major transitions. They were especially prevalent in the South because faith in self-treatment and the home healing arts was in harmony with tradition.

Many health-reform sects such as the Thomsonians, the homeopaths, and the hydropaths prescribed specific remedies and formulas. These prescriptions

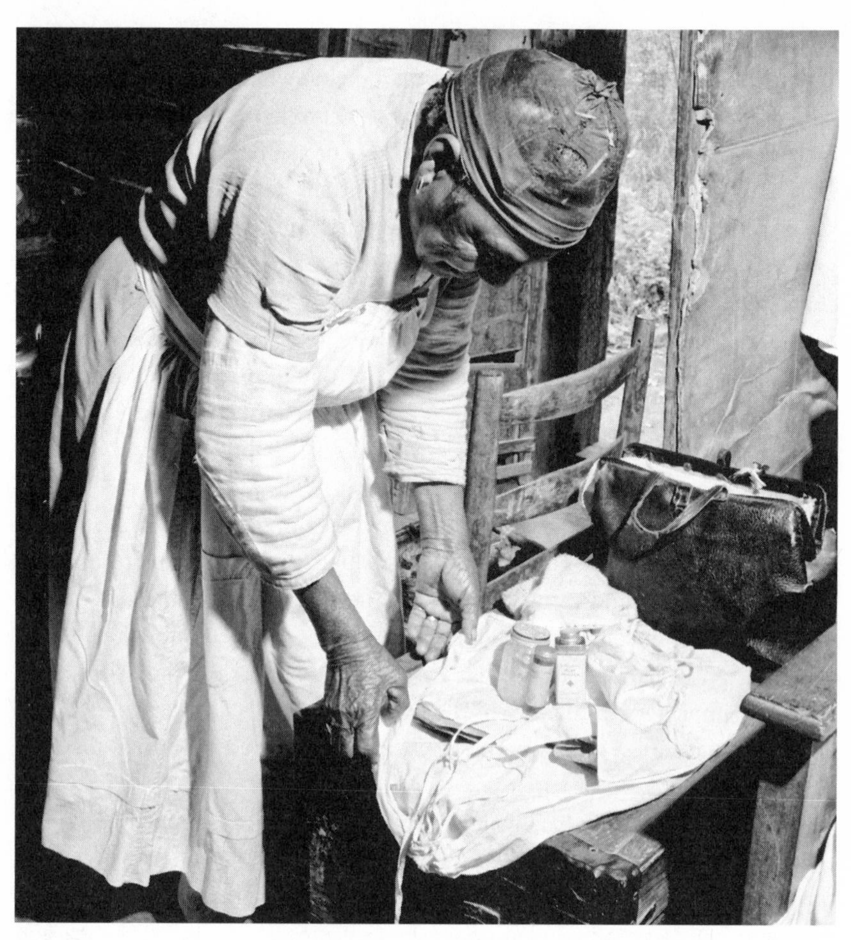

Midwife wrapping her kit to go on a call in Greene County, Ga., 1941 (Jack Delano, photographer, Library of Congress, [LC-USF34-046569-E], Washington, D.C.)

were added to the already rich base of healing knowledge in the South. Magical cures, formulas for healthful living, medicinal cures derived from local plants and animals, and professional health care coexisted with minimal conflict. Of the many remedies known in the South, some work well and some have proved to be useless or dangerous. Behind them are generations of women who grew and gathered plants, raised animals, treated injuries, applied poultices, administered medicines, and prepared barks, herbs, and plants for treatments. Many of these beliefs and practices, which point out the self-sufficiency and independence of southern women and their families, are preserved in the *Foxfire* books.

Midwifery was another special preserve of women. With the exception of such areas as Appalachia and the Ozarks, southern midwifery was an occupation dominated largely by black women who passed their skills and knowledge down to their mature daughters or nieces. Although the practices of midwives were allegedly a cause of childbed mortality rates higher than those of the North, more recent investigations suggest that other factors, especially nutrition, were responsible.

In the 1920s programs to train midwives in obstetrics were supported in southern states by welfare. Midwives met in the county health departments, and under the supervision of public health officials, nurses learned how to keep germs away during births, fill out birth certificates, and care for the equipment they used. By the 1970s nearly all traditional midwives were retired from practice, although nurse-midwifery training programs continue today.

Nurse-midwifery increased greatly in importance when Mary Breckinridge began the Frontier Nursing Service in the 1920s. By the 1960s teams of medically supervised nurse-midwives from the University of Mississippi worked in locations throughout the South where infant deaths were disproportionately high. These nurse-midwives demonstrated that they reduced infant deaths and provided high-quality care.

In recent years the number and availability of physicians and other health professionals have increased, but there is also a greater acceptance of women as healers in roles outside the family. Today women are entering the medical profession at the same rate as men. These changes reflect an increasing involvement of women in healing roles beyond the home and an accompanying belief that women can share economic responsibility without neglecting their traditional roles in healing and caring for their own sick.

MOLLY C. DOUGHERTY
University of Florida

Amanda Carson Banks, *Birth Chairs, Midwives, and Medicine* (1999); Mary Breckinridge, *Wide Neighborhoods: A Story of the Frontier Nursing Service* (1952); Elisabeth Brooke, *Women Healers: Portraits of Herbalists, Physicians, and Midwives* (1995); Marie Campbell, *Folks Do Get Born* (1946); Alex Freeman, *Kentucky Folklore Record* (October 1974); Paul F. Gillespie, ed., *Foxfire 7* (1982); Guenter Risse, Ronald Numbers, and Judith Leavitt, eds., *Medicine without Doctors: Home Health Care in American History* (1977); Sharon A. Sharp, *Women's Studies International Forum* (October 1986); Karen Shelley and Raymond Evans, in *Appalachia/America: Proceedings of the 1980 Appalachian Studies Conference*, ed. Wilson Somerville (1981); Jack Solomon and Olivia Solomon, compilers, *Cracklin' Bread and Asfidity: Folk Recipes and Remedies* (1979); Wilbur Watson, ed., *Black Folk Medicine: The Therapeutic Sig-*

nificance of Faith and Trust (1984); Richard W. Wertz and Dorothy C. Wertz, *Lying-In: A History of Childbirth in America* (1989).

Health, African American

In the post–Civil War era, white southern physicians noted the rise of disease among newly emancipated blacks and predicted the eventual extinction of blacks from the United States despite all that medicine could do to prevent it. Many white laymen in the South held similar views. In fact, issues of black health have always concerned southern whites. Interest in the subject constitutes a minor theme in the region's history. It is paradoxical that so politically voiceless a group as southern blacks has received so much attention with regard to health, medical care, and disease characteristics. Important reasons for this trend exist and are rooted in southern medical, racial, and interregional history. Indeed, because southern whites for so long either subjugated or segregated blacks, they have never been able to escape taking at least partial responsibility for black health. Additionally, as race has been important in the South and in relations between the South and the rest of the nation, whites have used black medical distinctiveness, real and imagined, for political purposes.

Blacks brought with them from Africa to the New World a heritage of disease and medical care differing in many ways from that of the dominant Euro-American society. Some African health problems (e.g., sleeping sickness) and practices (use of certain herbs and tribal rites) disappeared for environmental reasons. Yet genetic maladies (sickle-cell anemia), disease susceptibilities (tuberculosis, respiratory infections) and resistances (malaria, yellow fever), medical treatments (herbs, voodoo), and medical theories (supernatural causes of disease) remained, influencing the lives of both blacks and their white neighbors. White medicine dominated, but slaves—ostensibly not the owners of their bodies—could and did choose to invoke, covertly when necessary, their own medical systems. Some white masters condoned or reluctantly accepted these subtle statements of black independence because they saw no alternatives, felt their slaves should enjoy some freedoms, or noted that the black healing approaches worked. Some of these practices and remedies passed from the black to the white medical world.

Various factors influenced the health status of African Americans during slavery. Living conditions played an important role because plantation slave quarters, though located in rural areas, developed many public health characteristics of a village or small town. Infectious disease epidemics, parasitic infestations, and human and other organic-waste disposal problems resulted from life in slave communities. Nutritionally unbalanced or nutrient-deficient food caused, at the least, weakness and, at worst, death from malnutrition or infection. Hard, unsafe working and living conditions; whipping and other physical punishments; pregnancy, childbirth, and gynecological problems among physically laboring women; and the psychological stresses of a life of servitude also contributed to poor health among slaves.

As African American scholar and social worker Vanessa Jackson notes, doctors also "discovered" several mental pathologies specifically attributed to African Americans during the 19th century, including "negritude" (a form of leprosy in which the victims' only cure was to become white), "drapetomia" (the mental "disease" that compelled slaves to flee from forced servitude), and "dysaethesia aethiopica" (a disease characterized by lethargy and dull mental faculties). The two latter disorders, identified by slavery proponent Dr. Samuel Cartwright of Louisiana, were thought to respond favorably to whipping. Additionally, in both the antebellum and postbellum periods of the 19th century, some considered a rise in cases of mental illness among free African Americans evidence of the adverse effect of emancipation on the race.

When slaves needed medical attention, owners usually insisted blacks turn to the Euro-American curing system rather than caring for themselves. Most states required that masters provide for proper maintenance of their human chattel. Some fulfilled this responsibility better than others, depending on widely varying factors including personality, number of slaves, financial status, master-slave relationship, presence or absence of owner, and threat of an epidemic to property (slaves) and to white lives. As a general rule, masters, mistresses, overseers, or overseers' wives used their own medical knowledge and skills on ailing slaves first; then, if these resources failed to reverse the course of disease or injury, they called the local physician. Whites handled illnesses within their own families in a similar manner, applying home remedies before calling in the physician, with one difference. They generally sought professional help sooner for white family members than for slaves.

As southern whites increasingly cared for blacks, observed how slaves fared in the southern disease environment, and recognized the efficacy of various medical treatments on bondsmen, they began to notice trends. Blacks seemed much more prone to some diseases, less susceptible or even immune to others, and more or less responsive to one or another remedy. A belief developed, taking hold between 1830 and 1860, that the South possessed different health conditions from the North or West, that southern medical care differed from that in the North, that blacks were medically different from whites, and that southern physicians were uniquely suited and trained to handle the region's health problems. Blacks, they argued further, were medically suited for the labor and environment of the South. With mounting antislavery pressure from the North, these observations became political points in the proslavery argument. However, no southern physician ever proved any of these ideas.

Emancipation in 1865 ostensibly released southern whites from the responsibility of providing medical care to blacks. Freedmen had to rely on their own resources for the basic necessities of food, clothing, housing, and health. In reality, however, blacks generally did not have the means to survive without help. This seemed to justify white predictions that black emancipation doomed the race, as whites perceived African Americans to possess an innate physical inferiority and dissolute lifestyle that made them dependent on whites. Black health did decline, the result of poor housing, malnourishment, overcrowding, and the psychological stresses of a new environment and lifestyle. Whites noted rising rates of tuberculosis, syphilis, and insanity among the African American population.

Federal intervention offered little hope, as the medical department of the Freedmen's Bureau brought only fleeting and unreliable assistance for blacks. The bureau's goal was to ease the transition from slavery to freedom for both blacks and whites. Though it employed physicians in key areas to provide hospital and dispensary care, the bureau failed in its medical mission on two counts. It established no mechanisms whereby blacks could either continue to receive medical care or learn to care for themselves. When the bureau left a state (1868–71), blacks were left to their own devices or to the largess of local white authorities and white physicians. Once again, whites took on responsibility for black health, at least in part. Some labor and shareholding contracts contained provisions for medical care paid for by the white landowner. City and county government authorities worried about the spread of infectious diseases from blacks to whites, especially as the black population of cities and towns rose and their physical movements became less restricted. By the 1890s, the larger urban areas established hospitals and clinics to care for both black and white poor. Some northern medical schools and several newly founded black schools (e.g., Meharry, Howard, Leonard) produced a small number of black physicians, many of whom opened practices in southern black communities, but the health and medical care of African Americans remained at a lower level than that of whites.

As long as black health was primarily a southern problem, the nation as a whole ignored or remained ignorant of the black's plight. But the post–World War I migration of rural southern blacks in search of better jobs and living conditions to northern and border state cities altered national perception. Federal, state, and local governments, as well as private sources, including the

Rosenwald and Rockefeller foundations, began to provide funding to study and remedy the poor health situation of African Americans. These institutional benefactors supported public health campaigns (the annual National Negro Health Week), the building of hospitals, medical education for blacks, health improvement programs (tuberculosis, child and maternal health, syphilis [including the notorious Tuskegee Syphilis Study]), and overall assessments of black health status and needs (such as Gunnar Myrdal's *An American Dilemma: The Negro Problem and Modern Democracy* [1944]). The result of all this activity was the desegregation of health facilities, a better awareness of black health problems, government assistance to pay for the high cost of medical care, a reduction in black mortality rates, and an increase in the number of blacks in the healing professions.

In the first decade of the 21st century, researchers noted that African American populations in the South were particularly vulnerable to stroke, diabetes, heart disease, HIV/AIDS, and infant mortality. The region has seen the greatest increases in its black population in recent decades, and the 2010 census revealed that more than half of the nation's African Americans currently reside in the South. Recent studies by the U.S. Department of Health and Human Services show that health disparities between African Americans and other racial groups continue to be significant. For example, in 2007 the life expectancy of an average American was 77.9 years, but for African Americans that average dropped to 73.6 (compared to a life expectancy of 78.4 years for the average white American). Another staggering statistic is the infant mortality rate. For the years 2004–6, the infant mortality rate in the United States for African Americans was 13.5 (infant deaths per 1,000 live births) as compared to 6.8 for all races and 5.7 for whites. As southern blacks continue to suffer a lower quality of health, where state and local government funds are inadequate to pay for black health needs, rural life and traditional African American healing practices persist, and blacks often cannot afford private medical care, the South continues to be a critical arena for black health issues in America.

TODD L. SAVITT
East Carolina University

KATHRYN RADISHOFSKI
University of Mississippi

Gaines M. Foster, *Journal of Southern History* (August 1982); James H. Jones, *Bad Blood: The Tuskegee Syphilis Experiment* (1981); *Journal of the National Medical Association* (numerous articles on the history of blacks in medicine in issues from the 1950s to the present); Kenneth F. Kiple and Virginia Himmelsteib King, *Another Di-*

mension to the Black Diaspora: Diet, Disease, and Racism (1981); Herbert M. Morais, *The History of the Negro in Medicine* (1967); Todd L. Savitt, *Medicine and Slavery: The Diseases and Health Care of Blacks in Antebellum Virginia* (1978), *Race and Medicine in Nineteenth- and Twentieth-Century America* (2007); Susan L. Smith, *Sick and Tired of Being Sick and Tired: Black Women's Health Activism in America, 1890–1950* (1995).

Health, Mental

Before the mid-19th century, there was little publicly supported mental health treatment in the South. By 1825, Virginia, which made the first public attempt to treat the insane before the Revolution, was the only southern state to have a hospital for the insane. The South lagged behind the rest of the nation in this regard: eight other asylums existed in states outside the South. The sufferings of the insane in the South, as elsewhere in the nation, were looked upon as the natural consequences of a stern, unbending Providence, meting out judgment to the wicked and the innately inferior. The shame brought on by such a concept bred an attitude of contempt for, and lack of interest in, the needs of the insane. The families that could afford special accommodations provided strong rooms in attics and barns to shut away the family shame, or they sent the insane member to a neighboring state where institutional care could be purchased. The dependent insane who were not considered violent were allowed to wander through the town begging for food and becoming the butts of children's ridicule. Only those who were considered dangerous to the public welfare or who were a nuisance to the community received any public attention. Motivated by fear, communities used the local jail or almshouse as the common solution to the problem of public protection from the violent.

The South's concern for the insane was awakened by the reform movement of 1825 to 1860. In that period, South Carolina, Georgia, Alabama, Louisiana, Tennessee, Missouri, North Carolina, Mississippi, and Texas opened the doors of mental health care to the indigent insane, radically altering the character of the mental hospital movement in the South and bringing it up to par with the rest of the nation. In the early stages of development, moral therapy was employed as the accepted mode of treatment in the new state hospitals. Moral treatment involved removing the patient from the community to an asylum, where therapy of kindness and consideration for physical and emotional needs would lead to a cure. The assumption was that the insane could be cured in institutions removed from local conditions that prompted the onset of insanity. Before the growth of large public mental institutions, the insane had been embarrassments to their families, but they had been curiosities to the public. Now removed from the community, the mentally disturbed no longer posed a public embarrassment or a threat to the community, but they were still a public spectacle. For instance, the transfer of patients to the new North Carolina Western Insane Asylum at Morganton created a circus atmosphere in the town when the residents lined the road to watch patients being marched from the train station to the hospital. Likewise, the constant urging by the superintendent at Dorothea Dix Insane Asylum for construction of a fence around the Raleigh, N.C., facility was not for the purpose of protecting the citizens of the town, but rather to control the townspeople who came to the hospital grounds to watch and generally excite the patients.

Although widely heralded in the United States as an effective and successful therapeutic method during the first half of the 19th century, moral therapy fell into disrepute before the end of the century. The failure of moral therapy can be attributed, in large measure, to the exuberance of superintendents who issued reports of high recovery rates to stimulate the founding of new mental institutions and the expansion of existing ones. Superintendents willingly squeezed every patient they could into the hospital. Overcrowding and inadequate financial support made moral treatment impossible to practice. Nevertheless, outside pressures continued to exist to transfer mental patients away from the local community to the central state hospitals. The result of overcrowding and the absence of adequate medical treatment was the creation of warehousing facilities where patients were put out of sight and, therefore, out of mind. No one found it necessary to deal with the profoundly negative attitudes toward mental illness that permeated society.

As the state became increasingly responsible, local government and, more importantly, individual families began to assume that mental illness was not their responsibility alone. Unfortunately, those operating local hospitals were not attuned to the dangers of relegating responsibility for the mentally ill to a central state hospital. A relatively secure and simple hospital routine provided for a patient enabled that patient to avoid facing the more complex problem of life "on the outside" and created, more often than not, a pathological dependence on the institution. Recognition of this particular problem prompted the movement toward community clinics and local mental health programs.

Community responsibility was encouraged after World War II when three major factors combined to reverse the pressures on large state hospitals: the introduction of psychotropic drugs, a development of wartime research; federal support for research and mental health centers, prompted by the reports of various presidential and congressional commissions in the 1960s; and civil rights legislation and Supreme Court decisions on behalf of mental patients between 1961 and 1975 that dramatically changed state hospital census patterns. Between 1955 and 1977, patient enrollment in mental hospitals declined from more than 500,000 to less than 200,000. In the same period more than 800 community mental health centers were established. The psychiatric patient was returned from the large state hospital to his or her home community. Between 1955 and 1975, psychiatric patient care in state hospitals declined by 50 percent. Outpatient care in community clinics increased 70 percent in the same period.

The South has played a leading role in this movement. No southern state, though, has established a smooth transition from institutional care to community care. Meeting existing needs of the mentally disturbed at the community level rests on three factors: proper distribution of state resources, continuing public support of research and local community acceptance of mental health centers, and establishment of halfway houses and outpatient services. Every southern state has experienced a rapid growth in community mental health centers and comparative declines in state hospital populations, yet a commensurate shift in the allocation of funds to support local clinics has not occurred. The situation in Texas became typical of the resource distribution of all the states in the South. Between 1965 and 1977, Texas established 28 community centers serving 82 percent of the population of the state. Yet only 9 percent of the state's support for mental health went to community mental health centers. Although the situation has improved since 1959 in the area of research funding, when only four southern states (Florida, Louisiana, Tennessee, and Texas) were allocating more than $25,000 annually for research, every southern state commission cites the shortage of research funds as a deterrent to providing an adequate mental health system.

The late 20th and early 21st centuries continued to see a decline in state hospital populations, but further illustrating the shift away from institutionalizing the mentally ill and the difficulty in relying on community health centers to provide for their needs was the 2011 U.S. Department of Justice (DOJ) report that claimed many states across the South, including Virginia, Florida, North Carolina, Georgia, and Mississippi, were in violation of the Americans with Disabilities Act (1990) because they relied too heavily on mental health centers to institutionalize people with intellectual disabilities rather than providing the necessary support to live within their own communities. The DOJ report on the care Mississippi provides for the mentally ill states, "The state's reliance on institutional care harms residents of institutions. Not only are individuals segregated and denied the opportunity to participate in the everyday activities of community life, but they are subject to stigma and at risk of physical harm." By early 2012 many of these southern states had reached broad settlements with the Justice Department, but others were contesting the department's findings.

Hurricane Katrina, one of the most devastating and traumatic events in southern history, left as many as 2.5 million Gulf Coast residents displaced from their homes and more than 1,800 dead, including 1,577 in Louisiana and 231 in Mississippi. Many of those who survived the storm lost homes, friends, and family, and the stresses of coping with the loss of loved ones and the confusing, frustrating, and overwhelming process of rebuilding or relocating exacted a severe psychological toll. For several weeks after the storm, the Centers for Disease Control and Prevention surveyed residents of the New Orleans area for post-traumatic stress disorder and found that 50 percent of those surveyed indicated a possible need for mental health assistance, while 33 percent indicated a probable need for assistance. Illustrating the stresses associated with the event, estimates showed that in the first four months after Katrina suicides tripled and the murder rate increased by more than 38 percent in Orleans Parish. In Mississippi, crisis calls to Project Helpline, which chiefly deals with victims of depression and anxiety, reported an increase in calls by 61 percent between March and May 2006 as compared to October to December 2005, which indicates that rebuilding in Mississippi and Louisiana after Hurricane Katrina was, and continues to be, an exercise in rebuilding minds as well as communities.

CLARK R. CAHOW
Duke University

JAMES G. THOMAS JR.
University of Mississippi

Associated Press, "Department of Justice Critical of Mississippi Mental Health Programs" (29 December 2011), "Department of Justice Report Blasts Virginia's Mental Health Services" (11 February 2011); Leopold Bellak, ed., *A Concise Handbook of Community Psychiatry and Community Mental Health* (1974); Clark R. Cahow, *People, Patients, and Politics: A History of North Carolina Mental Hospitals, 1848–1960* (1982); Norman Dain, *Concepts of Insanity in the United States: 1789–1865* (1964); Gerald N. Grob, *Mental Institutions in America* (1973); Jim C. Nunnally, *Popular Concepts of Mental Health: Their Development and Change* (1961); J. Rabkin, *Schizophrenia Bulletin* (Fall 1974); Jean Rhodes et al., *American Journal of Orthopsychiatry* 80:2 (2010); R. Scott, *Schizophrenia Bulletin* (Fall 1974); Richard H. Weisler, James G. Barbee IV, and Mark H. Townsend, *JAMA: Journal of the American Medical Association* (2 August 2006).

Health, Public

The public health experience of the South, at least until the mid-20th century, was in many respects unique in the nation. Perceived as distinctive by northerners—and some southerners—for more than a century, the region's poor health record served as one more defining characteristic, one more peculiar burden added to southern history's extensive list. Although sharing many disease problems with the rest of the country, the South at various times exhibited maladies largely peculiar to itself—yellow fever in the 19th century and hookworm and pellagra in the early 20th century. Furthermore, certain infectious diseases that had afflicted the nation at large (e.g., malaria, typhoid fever, and tuberculosis) persisted at serious levels in the South until the 1930s and 1940s, years after having been brought under control elsewhere.

The "Sickly South" was an important facet of the region's image in the 19th century when yellow fever epidemics repeatedly ravaged the Gulf States and Lower Mississippi Valley. This "scourge of the South" attracted much negative attention, drained financial and human resources, deterred capital investment and urban population growth, and disrupted commerce and transportation. For much of the 19th century, state and local health measures concentrated on epidemic emergencies. With limited knowledge of the nature of diseases and modes of transmission, efforts at control through commercial quarantine and sporadic urban clean-up campaigns had little effect.

A turning point came in the 1870s and 1880s when germ theory and other medical advances brought increased understanding of disease processes. About the same time, the widespread yellow fever epidemic of 1878 led southern urban business interests to support increased public expenditures for such health-promoting, image-improving measures as public water supplies, drainage and sewerage systems, street paving, and garbage collection. These efforts clearly paid off in the improved state of health among whites in the urban South, although blacks showed slight improvements, as the new urban services rarely extended to poor neighborhoods.

The threat of yellow fever was finally brought under control through discovery of its transmission by the mosquito and the dramatic campaign against New Orleans's last epidemic in 1905. This demonstration of the power of "modern science" applied through the combined efforts of federal, state, and local health authorities, widely viewed as another turning point in southern health history, ended the long reign of "Yellow Jack" and removed what many called the last great obstacle to southern progress.

New obstacles soon appeared, however, as hookworm and pellagra were identified as prevalent ailments in the rural South. These peculiar debilitating

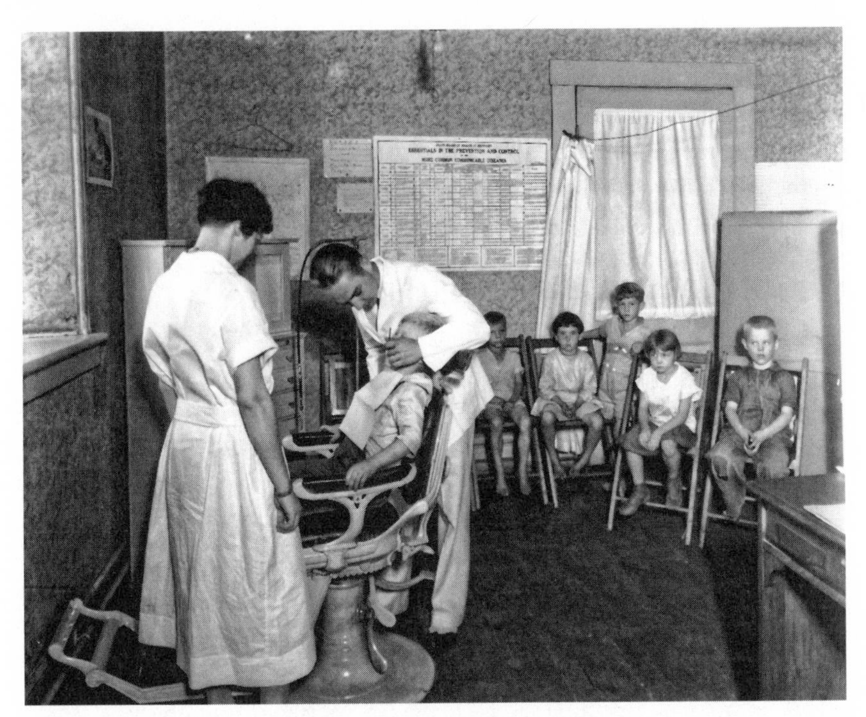

A Red Cross dental clinic in Kentucky, c. 1932 (Wolff, Gretter, Cusick, Hill Collection, Kentucky Historical Society, Frankfort)

disorders together with malaria, a persistent and widespread old malady, served to explain other long-standing features of the stereotyped South—its laziness, its backwardness—at least to the satisfaction of some "progressive" southerners and other Americans who sought in public health improvement a panacea for all the region's problems and a pathway to the modern world. State and local health authorities, assisted by the U.S. Public Health Service, the Red Cross, the Rockefeller Foundation, and other northern philanthropies, set forth to spread the gospel of health and bring modern medicine to the rural South during the next few decades.

Despite substantial achievements in developing health education and institutions, these efforts could provide only a palliative as long as basic conditions remained unchanged. Black and white southerners in the 1930s continued to manifest a remarkably high incidence of malaria, tuberculosis, typhoid fever, diphtheria, smallpox, venereal disease, hookworm, and pellagra, as well as high maternal and infant mortality rates.

Southern public health problems would not be solved by medical knowledge and health crusades alone; lasting solutions required broad social and economic change. Massive federal expenditures and changes associated with the Great Depression and World War II would finally transform the socioeconomic system, ending one-crop agriculture, stimulating urbanization and industrialization, and bringing about a higher standard of living for most of the southern population. With material improvement came the virtual disappearance in the postwar era of many diseases long sustained by the region's poverty. With the decline of nutritional-deficiency and infectious diseases, southern state and local health departments could devote more attention to chronic disorders, environmental and occupational health and safety, and other services. Nonetheless, prevention of communicable diseases remains a central part of public health vigilance.

Because of their severity, southern health problems have brought about the establishment of new public health institutions, local, state, and national. Louisiana created the first state board of health in the country in 1855 as a response to several widespread yellow fever epidemics. The epidemic of 1878, affecting the South and the Mississippi Valley interior and threatening the commerce of the nation at large, influenced Congress to establish a National Board of Health and, after that experiment failed, to assign an expanded federal role in quarantine and inspection service to the U.S. Marine Hospital Service, which became the Public Health Service by 1912.

Another distinctive health agency with a southern connection was the National Leprosarium in Carville, La., a state institution in the 1890s that became national in the 1920s until its closure in 1999. The earliest and most extensive development of county health departments (relying heavily on public health nurses) occurred in the South in the early decades of the 20th century, funded in part by the U.S. Public Health Service and the Rockefeller Foundation. Finally, the Centers for Disease Control and Prevention are located in Atlanta because of their origins in the Office for Malarial Control in War Areas. Established in 1942, in the center of the region where malaria was still most prevalent, the office sought to protect troops being concentrated and trained in the South, as well as the war industries' labor force. The agency's success in coordinating federal and state action led to its postwar continuation and expansion as the Contagious Disease Center, now the Centers for Disease Control and Prevention (CDC).

Today, Southerners suffer and die from the same major causes as the rest of the country—heart disease, cancer, stroke, and accidents—and they are served by similar state and local agencies. Some parts of the South still show the nation's highest death rates, infant mortality in particular, and a continuing high incidence of certain diseases—problems closely correlated with poverty and minority populations. Climate and the continued presence of appropriate mosquito vectors make the southern states still receptive to the threat of imported dengue, an infectious tropical disease, and yellow fever. Hence, while southern distinctiveness in health has been substantially diminished, it has not yet been eliminated altogether. New health concerns continue to appear across the South, even in the 21st century. One such health problem that has reached epidemic proportions is obesity, which contributes to other serious conditions such as diabetes and heart disease. In 2010, according to the CDC, 8 of the 12 most overweight states in the United States were from what is traditionally considered the South, the 12 being Mississippi (34.0 percent of the state's population), West Virginia (32.5), Alabama (32.2), South Carolina (31.5), Kentucky (31.3), Louisiana (31.0), Texas (31.0), Michigan (30.9), Tennessee (30.8), Missouri (30.5), Oklahoma (30.4), and Arkansas (30.1). Acknowledging the problem, the need for action, and the causes of the epidemic, Democratic state Rep. Steve Holland, chairman of the Mississippi Public Health Committee, said in 2007, "We've got a long way to go. We love fried chicken and fried anything and all the grease and fatback we can get in Mississippi." Because of southerners' predilection for fattening foods, a new law that year mandated 150 minutes of physical activity in public schools, and the Department of Education began phasing in restrictions on sodas and snacks in schools. As a federal response, in 2011 the Obama administration proclaimed September as being National Childhood Obesity Month, and Michelle Obama's "Let's Move!" initiative continued work with federal agencies and corporate and nonprofit sponsors to eliminate childhood obesity within a generation.

In 2009 the federal government addressed a need for increased health care among middle- and lower-class workers and families. Essentially, the United States Universal Health Care Act guarantees the majority of Americans health insurance coverage. One function of the new law is the expansion of Medicaid by subsidizing insurance premiums for households of those who earn less than 400 percent above poverty level. Consequently, as it stands in 2012, the new law would affect residents of the South more greatly than any other region's residents because the South is the poorest—and therefore the unhealthiest—region in the country. (In 2007, 9 of the 12 lowest-ranking household incomes in the country, according to the Department of Housing and Urban Development, were in the South.) In April 2012, the U.S. Supreme Court was deliberating the constitutionality of portions of the United States Universal Health Care Act.

Other organized efforts to improve southerners' health are underway across the region. For example, the nonprofit Southern AIDS Coalition (SAC) was established in 2001 as a result of the disproportionate federal support for southerners suffering with the disease and the overwhelming number of minorities afflicted with the disease across the region. The federally recognized organization works with state governments, community organizations and businesses, and those living with HIV/AIDS to "provide southerners with the opportunity to receive adequate HIV/AIDS prevention information, treatment, support services, and housing."

Despite the eradication of many diseases endemic to the South because of geography or economic conditions, the South still faces particular health challenges. Some are the result of the South's foodways culture and rural population, and others remain a result of economic conditions. Through continuing public health-care campaigns, the disparities between southerners' health and those in regions outside the South should continue to diminish.

JO ANN CARRIGAN
University of Nebraska at Omaha

JAMES G. THOMAS JR.
University of Mississippi

Associated Press, "Poverty Main Cause of Obesity Problem in South" (28 August 2007); Charles V. Chapin, *Report on State Public Health Work* (1915, 1977); CDC, *Morbidity and Mortality Weekly Report* (22 February 1980); John H. Ellis, *Bulletin of the History of Medicine* (May, August 1970); Elizabeth W. Etheridge, *The Butterfly Caste: A Social History of Pellagra in the South* (1972); Rebecca Ruiz, *Forbes* (17 November 2009); Susan L. Smith, *Sick and Tired of Being Sick and Tired: Black Women's Health Activism in America, 1890–1950* (1995); Dennis N. Tunnell, "Regional History of Southern Branch, American Public Health Association" (Ed.D. dissertation, University of Alabama, 1977); Margaret Warner, *Journal of Southern History* (August 1984).

Health, Rural

As of 2000, more than 40 percent of America's nearly 60 million rural residents lived in the South. Along with the Midwest, the region is the most rural in the country. Historically, the southern countryside and the city have differed in almost every way, including health. Although a national homogenization process changed much of rural culture and social structure during the 20th century, many non-metro counties still retain fundamentally distinct environments. Unfortunately, those in the South frequently exhibit the most execrable health indicators. Scholars have identified indigence, poor diet, inadequate housing, impure water supplies, lack of public transportation, and limited medical resources as key factors in explaining the overall health status of southern rural areas.

Of particular importance in understanding rural health is the problem of access, including the distribution of health services. Throughout rural areas there are shortages of physicians, dentists, and other health-care providers. Tellingly, in 2000, the Centers for Disease Control and Prevention (CDC) reported that both the South and Midwest had the lowest rates of HMO enrollment rates in the nation. In addition, poverty is noted as a crucial cause of poor rural health, especially in the South. Though poverty rates among non-metro residents decreased during the last half of the 20th century, the South retained 72 percent of American counties identified as persistently impoverished. As Perry B. Rogers states, "The level of family income in a population group is the most influential characteristic which determines whether a population will have health services which are appropriate and accessible."

Until the 1970s, public health activities provided the major health services in the rural South, with the exception of those areas serviced by private doctors. The Rural Sanitation Act of 1916 provided funds to improve such aspects of rural health as disposal of human waste, the protection of water supplies, and the control of insects. In the following decades, public health services continued to focus their efforts on malaria control, community sanitation, construction of sanitary privies, and sealing abandoned mines.

In 1935 the Public Health Services began to attack what it deemed the major problem facing rural families—the lack of adequate medical care. As a consequence, new programs for rural rehabilitation provided active medical-care personnel to reach more rural residents. After World War II, programs developed in the area of environmental health with a focus on communicable diseases. Increased institutional services were offered by newly constructed hospitals.

In the 1970s the U.S. Public Health Service launched a program called the Rural Health Initiative Projects. The program was designed to develop and systematize the delivery of health care in rural areas and included such projects as the National Health Service Corps, Community Health Centers, the Migrant Health Program, the Health Underserved Rural Areas Program, and the Appalachian Health Programs. These initiatives aimed to provide primary health care for rural areas, and resulted in the construction of local clinics in small towns throughout the South. Additionally, a significant number of rural hospital closures in the 1980s contributed to the establishment of the Federal Office of Rural Health Policy in 1987, a government institution similarly focused on facilitating access to health care among the inhabitants of non-metro areas.

Whatever is used to measure the health status of rural residents, one fact is clear: they remain worse off than any other population. Indeed, according to the CDC 2001 Urban and Rural Health Chartbook, rural inhabitants are more inclined to smoke and less likely to exercise and consume nutritional diets than their urban counterparts. Several predominantly rural groups have disproportionately severe health needs, including southern rural blacks, Latinos, Appalachian and Ozark whites, aged migrant workers, illegal aliens, and residents of environmentally polluted areas. Specifically, in the late 1990s and early 2000s, researchers found that rural African Americans in the South had critically elevated levels of poverty and were much less likely to be insured than those living outside the region. Likewise, studies from this period indicated non-metro southern counties with large African American constituencies were significantly underdeveloped in regard to municipal infrastructure.

Research conducted over the past several decades revealed the epidemiological patterns of both blacks and whites in the rural South reflect serious health problems. The CDC 2001 Chartbook demonstrated that rates of infant mortality, mortality in general, ischemic heart disease, inactivity due to chronic disease, and death from unintentional injury are substantially higher for rural southerners than for the region's metropolitan residents. Proportionally, rural counties in the South showed the greatest incidence of stroke nationwide, and their inhabitants were the least likely to have seen a dentist in the past year. Additionally, in recent years the South's communities in general, and its rural populations in particular, demonstrate an increased risk for contracting AIDS.

Obesity has also emerged as a salient health scourge in the region, as the South claimed 8 out of the 12 most overweight U.S. states in 2010. This fact is not particularly surprising considering that in 1998 the counties with the highest proportion of their population remaining physically inactive during leisure time were found in the South, and the lack of leisure-time physical activity has continued to decrease in the years since. Likewise, studies show that both rurality and residence in the South put children at high risk for obesity. Scholars identify food deserts (locales with a paucity of traditional edible resource outlets) as a significant contributing factor to rural southern obesity. Because both impoverished and non-metro neighborhoods often lack standard supermarkets, residents in these areas shop primarily at convenience stores, where processed, high-calorie, nutrient-poor food dominates and produce is rare and overpriced. In 2007 the Southern Rural Access Program (SRAP) responded to the crisis, launching the annual Southern Obesity Summit. The summit joined other SRAP initiatives and the Appalachian Regional Commission in developing region-specific tactics to combat deleterious health trends in the South.

The rural southern poor, both black and white, have been known to utilize folk medicine and share information about how to alleviate health problems. Middle-class whites often use folk remedies as well but do not use folk healers as frequently as poor people do. Because of limited access to health services, a unique rural culture emerged in the region, aimed at not only explaining illness but also offering ways to heal the sick. These folkways complement the scientific medical system and do not preclude the use of medical services. The poor in rural areas often share a sense of community, a worldview that is reflected in their health ideas, many of which are inextricably bound to religious beliefs. Likewise, southern rural people are wont to turn to kith and kin in times of illness and misfortune. Changes have occurred in rural areas, but traditional ideas of health remain in the face of policy initiatives that have gradually increased health services for non-metro populations over the past century.

CAROLE E. HILL
Georgia State University

KATHRYN RADISHOFSKI
University of Mississippi

M. C. Ahearn, U.S. Department of Agriculture, *Agriculture Information Bulletin* No. 428 (1979); C. L. Beal and Glenn V. Fuguitt, in *Social Demography*, ed. K. L. Taeuber, Larry L. Bumpass, and James A. Sweet (1978); James H. Copp, *Rural Sociology* (December 1972); Karen Davis and Rau Marshall, *Research in Health Economics* (1979); J. Lynn England, Eugene Gibbons, and Barry Johnson, *Rural Sociology* (Spring 1979); Dorothy M. Gilford, *Rural America in Passage* (1981); Nina Glasgow, Lois Wright Morton, and Nan E. Johnson, *Critical Issues in Rural Health* (2004); Carole E. Hill, *Current Anthropology* (June 1977); Olaf F. Larson, in *Rural U.S.A.: Persistence and Change*, ed. Thomas R. Ford (1977); Holly Mathews and Carole E. Hill, *Perspectives on the American South*, vol. 1, ed. Merle Black and John Shelton Reed (1981); Peter A. Morrison and Judith P. Wheeler, *Population Bulletin* (October 1976); Thomas C. Ricketts, *Rural Health in the United States* (1999).

Health, Worker

No concept of occupational health existed in the South or elsewhere in America until after 1910. "It is well known that there is no industrial hygiene in the United States," a Belgian labor expert told the International Congress on Occupational Accidents and Diseases in Brussels that year. In the next decade, however, widespread interest in social justice bore fruit in the recognition of hazards in the workplace. The U.S. Public Health Service (PHS) was particularly active, and the South was the site of one of its early efforts.

The first disease associated with southern industry was pellagra, a dietary deficiency disease that affected not only southern mill workers but tenant farmers, sharecroppers, and the poor wherever they lived. It was a peculiar and often fatal malady, marking victims with a distinctive rash and sometimes leaving them insane. The disease lost much of its mystery when Dr. Joseph Goldberger of the U.S. Public Health Service proved that pellagra was caused by an inadequate diet, a product of both the peculiar dietary habits of the region and the poor economic conditions under which many southerners lived. The fatback, cornbread, and syrup diet consumed three times a day by tenant farmers and mill workers was a vestige of the frontier past when settlers depended on corn, pork, and cane for their food. Gripped by tradition, many southerners clung to this diet long after the frontier was gone. Their choice was reinforced by economic considerations. Wages, traditionally low in the southern mills, fell even lower in the fall of 1920, and Goldberger predicted a pellagra epidemic. In a brilliant epidemiological study of seven mill villages in South Carolina, Goldberger and his associates had already conclusively linked the incidence of pellagra to poor economic conditions. Irate defenders of the South denounced him and the PHS and refused all offers of aid. Despite their protests, pellagra did increase throughout the 1920s and did not vanish until a quarter century later, when scientists identified niacin as the missing factor in the diet. A greatly changed economy and an agricultural revolution made southern workers more prosperous and their diets more balanced.

The immediate hazards of the workplace were brought dramatically to the public's attention in the early 1930s by an incident at Gauley Bridge, W.Va. White and black laborers from mountain districts of the South were brought to West Virginia to dig a waterpower tunnel through pure sandstone and quartz. The work began in 1929, and by the time it was completed three years later, 500 men had died of silicosis, pneumonia, and tuberculosis. Some dropped dead on the job and were buried within hours, sometimes two or three men in a single grave. Sensational compensation cases were tried in the courts, and Gauley Bridge became a symbol of danger on the job, particularly from dust.

As a result of this tragedy, many commissions were formed to study the relationship of dust to health, though some of them may have been more concerned with forestalling massive claims than with improving working conditions. A survey of Virginia industries in 1938 showed that industrial officials believed that if their factories were free of silica dust, they were free of all occupational disease hazards, but as the Virginia study showed, more than a quarter million workers, or one-tenth of the population of the state, were employed in industries where occupational diseases were known to exist. The greatest numbers were exposed to dust of one sort or another, prolonged exposure to which caused trouble in the upper respiratory tract.

The number of southern workers who became ill as a result of industrial hazards was difficult to ascertain. By 1944 at least six southern states required physicians to report occupational diseases—Alabama, Arkansas, Georgia, Louisiana, Mississippi, and South Carolina—but this requirement did not provide a satisfactory method of getting statistics. The laws were not standardized, nor was there any one agency in each state to whom physicians reported. In 1951, when four southern states participated in a pilot study to report occupational diseases, South Carolina found no cases of disease caused by dust in spite of its large textile industry. Nearly all states limited workmen's compensation claims to personal injury by accident, excluding occupational diseases altogether.

The first such disease to attract public attention was black lung, or pneumoconiosis, in the mid-1960s. A prevalence study showed that 1 of 10 miners working in the bituminous coal mines of Appalachia showed radiographic evidence of black lung, a disease marked by black spots on the lungs and greatly impaired breathing. One in five nonworking miners was affected. The study included miners in the southern states of Virginia, Tennessee, and Alabama and refuted the assumption that only miners of anthracite coal were subject to black lung. The disease probably had existed for many years before it became an issue. The danger that dust posed to health was increased with the introduction of mechanical loading equipment in the 1930s. These "man killers" greatly increased the dust level.

For 20 years the United Mine Workers promoted mechanization of the mines, believing that this would lead to higher wages and economic security, but the work force shrank steadily, and the dust, noise, and other hazards increased. The black lung revolt of 1968 was triggered by the refusal of the West Virginia legislature to make the disease compensable under state law. A strike took 40,000 miners off the job, and a violent explosion in a Farmington, W.Va., mine killed 78 men, setting off a national political debate on black lung disease and resulting in the passage of the Coal Mine Health and Safety Act of 1969. This bill detailed mandatory work practices in the industry and provided compensation for victims of black lung and the widows of those who died from the disease. Activists among the miners contended that the ultimate cause of the disease was economic: mine owners did not spend enough to keep down the dust.

The increasing role of the federal government in protecting workers' health climaxed in the passage of the Occupational Safety and Health Act (OSHA) of 1970. A year later the first health standard was set, a temporary one for asbestos. This standard grew out of a long-range study of asbestos workers in Tyler, Tex., which showed that 39 percent of workers with more than 10 years' employment in the company had asbestosis and that 30 percent of those workers who had massive exposure to asbestos fibers would develop cancer. The disclosures were so sensational that the Pittsburgh Corning Corporation, which owned the factory, closed it and buried most of the equipment.

Before the enactment of OSHA, the textile industry was barely aware of problems with cotton dust. Industrial air-conditioning improved the mill environment in the 1950s, but in the next decade speculation grew that raw cotton dust endangered workers' health. It was found to cause byssinosis, or brown lung disease, which is marked by chest tightness, shortness of breath, coughing, and wheezing. The disease begins with "Monday fever," when workers with long-term exposure to cotton dust fall ill every Monday. Later, symptoms last over several days, and finally the disease becomes chronic—the worker is disabled and the effects are irreversible. Industries responded to public pressure and stepped up the installation of dust-cleaning equipment in the 1970s. OSHA set standards for cotton dust exposure in 1978 that were upheld by the U.S. Supreme Court in 1981. As a result of the new standards, fewer than 20 byssinosis-induced deaths occurred across the United States annually between 1979 and 1999. Fifty percent of those deaths, however, occurred in Georgia and the Carolinas where the textile industry predominates.

The manufacture of any new product may threaten workers' health. Viscose rayon plants in Virginia, North Carolina, and Tennessee were among those that changed their methods of manufacture in 1937 after it was found that workers were dangerously exposed to carbon disulphide, a poison that affects the central nervous system, causing paralysis in the legs or manic-depressive insanity. More recently, the increase in the number of available chemical compounds has multiplied the danger to workers. The kepone case in Hopewell, Va., in 1976 is illustrative. Workers making insecticides from kepone developed nervous disorders, bodily shakes, and sterility after exposure to the product. Waste from the plant, emptied into the James River, poisoned the river valley, and wiped out fishing there for years.

Labor unions have been important in the struggle for a safer workplace. The textile workers union forced OSHA to establish standards for cotton dust and called for programs of medical surveillance and wage retention for workers too sick to work in dusty areas. Their emphasis has been on forcing industry to modernize, not only to improve working conditions for labor but to make the companies economically viable in a competitive world.

One recent cause for working-conditions concern was the cleanup after Hurricane Katrina. In 2005 New Orleans fire and police departments requested that the National Institute for Occupational Safety and Health (NIOSH) assist them by determining the health risks that emergency responders faced. Some of the health threats NIOSH recognized were the handling of the deceased, thermal stress, exposure to elevated amounts of carbon monoxide, and infection. Another potential risk was the exposure to toxic and biological contaminants encountered when wading through hip-high floodwaters. Much like those who worked at Ground Zero following 9/11, relief workers also faced the threat of post-traumatic stress disorder.

Cleanup of the Gulf Coast after the 2010 Deepwater Horizon oil spill in the Gulf of Mexico contained its own health risks, especially for those who were exposed to chemical dispersants used to break up the oil. Workers complained of having flulike symptoms such as respiratory problems, nausea, and headaches—complaints similar to those of people who worked on the *Exxon Valdez* oil spill in Alaska in 1989. According to the Centers for Disease Control and Prevention, long-term exposure to dispersants can cause central nervous system problems or do damage to blood, kidneys, or liver. The Gulf Long-Term Follow-Up Study, conducted by the National Institute for Occupational Safety and Health, began in March 2011 and intends to survey 55,000 cleanup workers through the year 2021. Although the long-term effects of the spill will not be conclusive for many years, a common opinion held by the scientific community and Gulf Coast residents alike is that the health risks resulting from the oil spill are liable to last for generations.

ELIZABETH W. ETHERIDGE
Longwood College

JAMES G. THOMAS JR.
University of Mississippi

Daniel M. Berman, *Death on the Job: Occupational Health and Safety Struggles in the United States* (1978); Paul Brodeur, *Expendable Americans* (1974); Elizabeth W. Etheridge, *The Butterfly Caste: A Social History of Pellagra in the South* (1972); Joseph Goldberger, George A. Wheeler, Edgar Sydenstricker, and W. I. King, *Hygienic Laboratory Bulletin No. 153* (1929); Jeremy P. Jacobs, *New York Times* (21 April 2011); John Manuel, *Environmental Health Perspectives* (January 2006); Joseph G. Montalvo Jr., ed., *American Chemical Society Symposium Series No. 189* (1982); National Institute for Occupational Safety and Health, Worker Health Chartbook (2004); George Rosin, *Preventive Medicine in the United States, 1900–1975: Trends and Interpretations* (1975); Barbara Ellen Smith, in *Health and Work under Capitalism: An International Perspective*, ed. Vincent Navarro and Daniel M. Berman (1983); Renee Twombly, *Environmental Health Perspectives* (January 2006).

Maternal and Child Health, Urban

Maternal and child health status provides a snapshot of an urban community's social and economic vibrancy. Improvements in maternal and child health benefit society not only by increasing survival rates but in improving the overall quality of life for families and communities and, thus, nations. From a population-health perspective, researchers who link health to economic development argue that interventions to improve maternal and childhood health are a social investment that can have positive long-term economic effects as birthrates adjust, healthy children are better prepared to be educated and be productive citizens, and healthier populations require less to be spent on costly medical care. Maternal health and child health conventionally address a broad continuum of topics, including health issues related to preconception, pregnancy, delivery, postpartum, birth spacing, and early childhood. Highlighted here are three current critical urban maternal and child health issues: infant mortality, unplanned pregnancy as it links to preconception health, and sexually transmitted infections (STIs). These issues are important in the South, the region reporting high rates of infant mortality, teen pregnancy, and sexually transmitted infections. Given the direct relationship between population health and economic development, the potential economic and community development in the South will continue to be constrained, especially for minority and urban poor subpopulations.

The most troubling health issue afflicting southern maternal and child well-being is the phenomenon of infant mortality—defined as an unexplained death in a child up to three years old. Infants born in the South are more likely to be born prematurely, to be low weight, or die than infants in other parts of the United States. Moreover, according to the Centers for Disease Control and Prevention (CDC), the incidence of infant mortality within the Mid-South (west Tennessee, Mississippi, east Arkansas) rivals rates experienced in developing nations. The CDC indicates that the U.S. infant mortality rate was 6.86 infant deaths per 1,000 live births in 2005; in Tennessee the rate was 8.77 in the same year. In Memphis, Tenn., the rates of infant mortality, according to the Memphis and Shelby County Health Department, showed an overall infant death rate of 13.8 per 1,000 live births in 2006 but the rate for African Americans was

19.0 per 1,000. The underlying reasons for this high rate are wide ranging and interrelated, including factors such as lack of early prenatal care, economic instability, and race. What is clear, however, is that low birth weight shares a high correlation with infant morbidity. Unfortunately, the means for preventing low birth weight in babies are imprecise, as solutions relate to both medical conditions and social/environmental factors, including the reduction of domestic violence and adolescent pregnancy. One solution favored by community health professionals is breastfeeding because it is a source of high-quality nutrition. The benefits of breastfeeding are great for both mother and child; studies now suggest a decreased risk of breast cancer in women who breastfed, and children that have been breastfed are less likely to be susceptible to autoimmune diseases. But breastfeeding rates vary by region. According to the CDC, southern states report the lowest percentage of breastfeeding at birth, six months, and one year. Expanded public education efforts to disseminate knowledge on the positive effects of breastfeeding for both mother and child are beginning to appear in the form of grassroots movements. Although social, ethnic, and cultural factors affect individual decisions to breastfeed, hospital practices also affect these decisions. Southern states score lower in practices supportive of breastfeeding when compared to other regions.

The community response to the infant mortality issues crosses sectors and includes a broad array of approaches and programs, such as nurse home visitation, promotion of early prenatal care, parenting education, domestic violence programs, faith-based family counseling, and even ways to safely abandon an unwanted infant (e.g., bringing the infant to a hospital or fire station, with no questions asked). Tennessee is supporting health promotion programs such as Centering Pregnancy (a group prenatal support program) as well as a March of Dimes–designed campaign in Shelby County, Tenn., Community Voice, to educate volunteers about infant mortality, especially as it is affecting the African American community, and to elicit their support in sharing what they have learned throughout the community. Southern states also work toward establishing formal, community-based organizations responsible for fetal and infant mortality review in urban areas in order to identify relevant factors and recommend system changes where needed.

According to recent data from the National Campaign to Prevent Teen and Unplanned Pregnancy, American women are experiencing approximately three million unplanned pregnancies each year; the concern is the high correlation between the lack of purposeful and proactive attention to preconception health and child morbidity. Preconception care is pivotal in eliminating many preventable conditions, including neural tube defects that can result in chronic ill health for the child. Preconception health is also important to assuring medical management of chronic conditions to reduce their potential adverse effect on pregnancy outcomes. The CDC indicates that in 2002 approximately 6 percent of adult women aged 18 to 44 years had asthma, 50 percent were overweight or obese, 3 percent had cardiac disease, 3 percent were hypertensive, 9 percent had diabetes, and 1 percent had thyroid disorder. The potential for preconception health care is greatly diminished among adolescent mothers, who typically have the highest rates of unplanned pregnancy. High rates of adolescent pregnancy, especially for teens lacking family financial resources, can be a serious negative factor affecting the long-term quality of life for both adolescent parents and their children; complications for childbearing women under the age of 20 greatly increase compared to women 10 to 15 years older. The National Campaign to Prevent Teen and Unplanned Pregnancy reports that southern states have among the highest rates of teen pregnancy, with the South having 9 out of 18 of highest teen birth rates in the nation and Mississippi (103) and Florida (97) having the highest southern teen birth rates. The community response to the unplanned pregnancy/preconception health issue is necessarily broad and includes health and wellness promotion, education, and awareness activities; promotion of early prenatal care; faith-based health outreach programs; family-life education in schools; and health campaigns targeting physical fitness, good nutrition, and health screenings.

Other types of disease potentially harmful to children result from the transmission of sexually transmitted infections, including HIV/AIDS. According to the CDC, the presence of HIV in pregnant women if treated properly through antiretroviral drugs can result in transmission rates to the fetus as low as 2 percent. The link between concentrated poverty, high incarceration rates of the urban poor, and substance abuse is thought to be a factor in STI rates in urban areas. STIS include the transmission of both bacterial infections, such as chlamydia and gonorrhea, and viral infections, such as genital herpes and genital human papillomavirus (HPV). STIS can increase the risk of preterm delivery and, if untreated, can cause cervical and other cancers, infertility, and other complications. In pregnant women, STIS potentially can be transmitted to the baby before, during, or after delivery. According to the CDC, almost 1.1 million chlamydial infections were reported nationally in 2006, and women infected with chlamydia are up to five times more likely to become infected with HIV, if exposed. Mississippi (745.1), South Carolina (611.7), and Alabama (546.9) rank three out of four of the highest national rates of reported cases of chlamydia in 2007. Testing during prenatal care and regular health screenings can help to prevent the harmful effects of STIS. The community response to this issue includes continued monitoring and follow-up of cases by local public health departments, health promotion activities, and health education about safe sex practices.

Maternal health and child health in urban settings occur within a social, political, environmental, and economic context. According to the Environmental Protection Agency, the developmental environment can determine various physiological parameters in the developing individual that can affect adult health status. Further, evidence shows that a female's in utero and early postnatal environments influence her own reproductive success later in life. Moreover, other urban-related socioeconomic factors place mothers' and children's health at risk, including urban neighborhood characteristics that directly relate to dietary risks and impede healthy physical activity; food insecurity linked to poverty that compromises nutritional status of pregnant women and developing children; lack of access to health providers; mental health stressors associated with urban-related social disorganization; poor habitat conditions that compromise well-being and healthy development; limited availability of public or other transportation that acts as a barrier to accessing health and social services; childcare responsibilities that constrain the mother's ability to comply with care appointments; domestic violence that tends to be greatest toward women in their reproductive years—violence experienced during pregnancy can result in numerous health complications for the child including low birth weight and even fetal death; and racial and economic disparities that constrain economic and social opportunities.

Immigrants in the United States create an additional set of challenges for those concerned about urban maternal and child health and health disparities. The growing presence of immigrants, especially young women and children in cities, in suburbs, and in rural areas of the South, raises new questions about social inequality and health care. Responding to different languages, cultural expectations, and lifestyles is a challenge for urban health providers and can further strain community resources.

Individual values, beliefs, and perceptions play an important role in maternal and child health. Some feminists argue that reproduction and sexual health have become too medicalized in the United States. Further, culture and ethnicity affect health beliefs about pregnancy, infant care, immunizations, and wellness and can influence maternal decisions about health behaviors and modalities of care for herself and her children, allopathic medicine, naturalistic health approaches, or folk medicine. Culture and ethnicity also affect an individual's level of trust of health-care providers and institutions and can affecting her willingness to seek health-care services.

Mothers and children have generally been viewed as having a legitimate claim for society's assistance. The degree of political support for that claim has, however, wavered over time and has involved debates about whether women who deviate from normative ideas about appropriate motherhood and sexuality deserve assistance. Historically, such debates have been especially strong in those southern states where officials have been willing to punish "promiscuous" behavior by withholding benefits. However, broad support has generally been sustained for maternal and child health and social services, as evidenced by numerous federal- and state-funded programs that are focused on children and pregnant women, such as Medicaid; Women, Infants, and Children (WIC); and Healthy Start. Political ideology directly affects the types of programs supported by communities: for example, comprehensive sexuality versus abstinence-only education programs for teens. Increasingly, public health advocates are calling for more participatory community-based approaches in order to engage the community as full partners in programs and to assure cultural sensitivity. Undoubtedly, a systems perspective that respects the effects of place is needed to understand the public health issues affecting southern maternal and child health status and to assess the public policy and community responses to these issues.

JOY A. CLAY
BRIDGETTE R. COLLIER
University of Memphis

Dolores Acevedo-Garcia, Theresa L. Osypuk, Nancy McArdle, and David R. Williams, *Health Affairs* (March–April 2008); Centers for Disease Control, www.cdc.gov; Ronald David, *Focus* (September–October 2005); Ruth Feldstein, *Motherhood in Black and White* (2000); Nicholas Freudenberg, Sandro Galea, and David Vlahov, eds., *Cities and the Health of the Public* (2006); Sandro Galea, Nicholas Freudenberg, and David Vlahov, *Social Sciences and Medicine* (2005); Linda J. Koenig et al., *American Journal of Public Health* (June 2006); Memphis and Shelby County Health Department, "Infant Mortality Statistics," www.shelbycountytn.gov; John Mirowsky, *Social Forces* (September 2002); David Mirvis and David Bloom, *JAMA* (2 July 2008); *Morbidity and Mortality Weekly Report* (21 April 2006); National Campaign to Prevent Teen and Unplanned Pregnancy, *Responding to the Increase in the Teen Birth Rate: Analysis from the National Campaign to Prevent Teen and Unplanned Pregnancy* (January 2009), www.thenationalcampaign.org/resources/birthdata/analysis.aspx.

Medical Care, Public Health, and Race

In the antebellum South, Americans were vulnerable to many diseases and conditions that scientific information and medical knowledge have rendered rare in the present-day United States. Without antibiotics, advanced surgical techniques, understandings of infection, vaccines, or effective sewage systems, antebellum Americans suffered from postpartum infection, tuberculosis, and a host of other problems. But African Americans, particularly southern slaves, had health issues that sprang directly from their enslavement. Among these were malnutrition, parasites, tuberculosis, and childbirth complications. Whippings, which were carried out across the South from the colonial era through the Civil War, injured skin, muscle, and occasionally internal organs. Slaves' labor was also frequently physically dangerous. Domestic work such as cooking and laundry carried risk of burns and other injuries. Agricultural labor exposed slaves to injuries from machinery and livestock.

Most African American slaves moved between two parallel medical systems. Over the years, slave communities developed their own system of medical treatments and understandings of health and illness. Slave practitioners who were largely illiterate received medical training from other slaves. Their treatments tended to rely heavily on local flora. For these healers, curing the body often intersected with the patient's spiritual and emotional health as well as the patient's relationship to family and community. Largely divorced from concerns of the market and labor production that informed white doctor's treatment, African American medicine was a key component of slave community and spiritual life.

Slave owners, on the other hand, wanted healthy slaves who would be able to labor productively or bring a high price on the market. Often masters relied on a mixture of patent medicines, homeopathy, and their own or an overseer's medical skill to treat a wide variety of chronic and acute conditions. Many masters, particularly when dealing with slaves "in their prime," spent considerable resources treating illness. Many called in white doctors, paid for medicines and surgeries, and asked one another and professionals for advice on keeping slaves healthy.

White doctors' interest in slaves extended beyond desires to cure them for either humanitarian or financial interest. White physicians in the South often used slave patients as research subjects and the remains of deceased slaves for autopsies and educational purposes. Although some of these experiments and procedures attempted to locate the physical essences and implications of racial difference, most focused on treatments that they hoped would apply to white patients. Unsurprisingly, white doctors virtually never asked slaves or their next of kin for permission to use dead bodies or to perform research on the living. White doctors could hire slaves with specific conditions to serve as research subjects.

The best known of these projects is the work of Dr. J Marion Sims of Alabama. Sims performed 30 vaginal surgeries on three slave women suffering from vesicovaginal fistulas, a debilitating and humiliating condition, between 1845 and 1849. Sims used no anesthetic during these operations and eventually perfected a surgical technique and invented instruments that became standard for vaginal exams and surgeries. Sims's work, and indeed most medical practice under slavery, was predicated on slavery's central tenet—that another person could own one person's body, as opposed to just their labor. When the Civil War and Emancipation put an end to bodily ownership, the systems of medical care that had developed under slavery changed as well. In 1862 President Lincoln signed the Militia Act that allowed black men, regardless of their status, to enlist in the Union army. Black men who served in the military received an initial physical exam from an army physician. These records reveal a great deal about the general health of African American men (northern and southern) in the mid-19th century. African American soldiers, both those who had been slaves and those who enlisted as free men, suffered higher mortality and received less care than their white counterparts, as the army assigned few trained physicians to African American regiments. Because proportionally fewer of them served in combat roles, their rate of battle wounds was lower than that of white soldiers. The reasons for their high mortality rates converge around various factors. For example, many rural African Americans had not been exposed to childhood contagious diseases and were vulnerable to these in the close military quarters. In addition, African American troops performed much of the army's dangerous and physically taxing labor, such as digging fortifications and clearing land. The poor quality of many of the African American troops' uniforms and rations also weakened their defenses against illness.

In addition to African American men on active duty in the military, thousands of African Americans—men, women, and children—escaped plantations and fled to refugee camps administered by the Union army. These people often arrived in poor health and found that the crowded conditions, poor sanitation, and scarcity of food and shelter rendered them vulnerable. They sought out medical treatment both from Union army doctors and from the volunteers and employees of Freedmen's Aid organizations who staffed the camps.

After Emancipation, the federal government created the Bureau of Refugees, Freedmen, and Abandoned Lands. Although the bureau did provide various forms of relief for freedpeople, as did a number of private agencies, from its inception it resisted providing free and complete medical care to African Americans. Despite this, the bureau established hospitals and dispensaries across the South. Some freedmen's hospitals provided leadership roles for African Americans. For example, Alexander Augusta, a black physician who had trained in Canada and served as a Union army surgeon, worked at the freedmen's hospital in Georgia and later at Freedmen's Hospital in Washington, D.C. The latter hospital, which was largely funded by African Americans, grew into part of Howard University's medical school.

Many African Americans were unable or unwilling to rely on Freedmen's Bureau hospitals and instead tried to negotiate medical care into their labor contracts with employers. Still others formed mutual aid societies that negotiated with physicians as a group to provide benefits for their members. Many of these groups engaged in philanthropy as well, providing medical care for indigent black people in their communities.

During Reconstruction and through the turn of the century, black and white philanthropists invested in black schools throughout the South. As part of this movement, religious groups, preexisting colleges, philanthropists, and even entrepreneurs founded medical and nursing schools. Medical education was largely unregulated during this time, and like institutions that catered to white students, African American medical schools were wildly uneven in terms of curriculum and quality. In 1909, in a major step toward regulating medical education and creating national standards, Abraham Flexner, commissioned by the Carnegie Foundation, made a tour of medical colleges in the United States and Canada, evaluating their quality. Flexner assigned a grade to each school and occasionally explicitly recommended that a school close. Even before state and medical board authorities put in formal regulation measures, a low mark from Flexner effectively crippled a school's attempt to raise funds and recruit students. Flexner gave low marks to all but three African American medical schools: Howard University Medical School in Washington, D.C., Meharry Medical College in Nashville, and Leonard Medical School in Raleigh. The Flexner Report did shut down substandard medical schools, but these schools had provided a basic medical education to African Americans who then went on to serve black patients. White medical schools that survived the Flexner report did not admit black students, and the American Medical Association remained segregated. In response to these conditions, a group of African American medical professionals formed the National Medical Association, which is ongoing.

In the midst of these developments in African American medical education and institution building, experimentation on African Americans' bodies continued. The Public Health Service organized and funded the most notorious of these experiments. The Tuskegee Syphilis Study took place in Macon County, Ala., home of the Tuskegee Institute. In this study, which ran from 1932 until 1972, African American men who had tested positive for syphilis were followed and studied for years. The aims of the study were to learn the effects of untreated latent stage syphilis, more specifically on black men. Over the years, under the direction of both black and white doctors and one black nurse, the men received what they thought was treatment, but in fact was aspirin and iron tonics. Men in the study were discouraged from seeking out the legitimate syphilis treatment and kept ineligible for the draft during World War II because the military demanded testing and treatment for syphilis for all inductees. After providing data for 12 academic papers in medical journals, the study ended in 1972 amid a flurry of bad publicity. President Clinton issued a formal apology for the study in 1997.

In 1946, 14 years into the Tuskegee study, the U.S. Congress passed the Hill-Burton Act, originally named the Hospital Survey and Construction Act. The act provided grants and loans for underserved areas to build medical infrastructure, mainly hospitals and nursing homes. Institutions that accepted the funds were obligated to provide a "reasonable volume" of services to people unable to pay. Hill-Burton funded the construction of hospitals all over the South, and although the original language in the bill stipulated that projects that the initiative funded could not discriminate on the basis of race, the funding of separate-but-equal facilities continued until 1963.

The years since 1963 have seen increasing desegregation of medical schools, nursing schools, and hospitals. However, demographic patterns and access still translate into racial imbalances in many institutions and medical specialties. African Americans in the 21st century still suffer disproportionately high rates of asthma, obesity, type-2 diabetes, infant mortality, and HIV infection. These statistics emerge from a constellation of causes. Poverty, lack of health insurance, lack of access to primary care, poor nutrition, and environmental factors all play a role in the illnesses that African Americans get, how those illnesses are treated, and the outcomes of that treatment. A marked disparity between the health of African Americans and white people has persisted since before the nation's founding and continues today.

GRETCHEN LONG
Williams College

Sharla Fett, *Working Cures: Healing Health and Power on Southern Slave Plantations* (2002); Margaret Humphreys, *Intensely Human: The Health of the Black Soldier in the American Civil War* (2008); Susan Reverby, *Examining Tuskegee: The Infamous Syphilis Study and Its Legacy* (2009); Todd L. Savitt, *Medicine and Slavery: The Diseases and Health Care of Blacks in Antebellum Virginia* (1978), *Race and Medicine in Nineteenth- and Early Twentieth-Century America* (2007); Maria Schwarz, *Birthing a Slave: Motherhood and Medicine in the Antebellum South* (2006); Karen Kruse Thomas, *Journal of Southern History* (November 2006); www.womenshealth.gov/minority-health/african-americans.

Medical Centers

In visiting southern urban centers, one often gets the impression that there is a hospital on every corner. While this may be more perception than reality because of the prominent locations of hospitals in southern cities, it could be argued that the state of medicine in the South has historically been more advanced than that for most other areas of the economy. Although there are major medical centers all over the nation, southern medical centers appear to play a much more important role in the economies and culture of their respective communities than do medical centers in other regions of the country. Few cities outside the South are as closely associated with their medical centers as are many southern cities, a situation often noted by those in the healthcare field. For health professionals in particular, it is hard to think of Atlanta without reference to the Centers for Disease Control and Prevention or Memphis without mention of St. Jude Children's Research Hospital or Houston without considering M. D. Anderson Cancer Center. Indeed, Nashville is often referred to as the "Silicon Valley" of healthcare because of the number of important health-related businesses it has spawned.

Although it is true that most of the nation's prestigious medical schools and urban medical centers are outside the South, it is noteworthy that southern cities actually have a proportionate share of health facilities, given the lag the region exhibits with regard to many other industries. The relative importance of health care in southern urban centers is perhaps explained by the fact that, although there are a lot of things that a community can do without, health care is not one of them. The growth of major urban centers could also be attributed to the historical lack of large cities outside of a few regional centers. Thus, the growth of medical centers in places as diverse as New Orleans, Memphis, and Louisville reflects their position as regional centers with few communities of any significant size in their sphere of influence. Although this situation is no longer the case as even secondary medical centers in the hinterlands have access to contemporary medical facilities, these southern urban medical centers were established at a time when there were few other options, and most have retained their regional significance.

The importance of the health-care industry to the South in general and to southern cities in particular is reflected in the fact that the proportion of the work force involved in health care is higher in the South than in the rest of the nation. Indeed, there are certain southern cities where the employment in health care and social services is notably higher than the average. This is particularly the case for middle-sized southern cities that may not have historically had the industrial diversity of an Atlanta or a Dallas. Thus, the South has "health-care towns" where, according to the 2000 census, the industry is inordinately dominant. These include Little Rock (26 percent higher health-care employment), Birmingham (11 percent higher), New Orleans (9 percent higher), Louisville (8 percent higher), and Mobile (6 percent higher). Others slightly higher in health-care employment include Memphis, Nashville, and San Antonio.

The importance of health care to southern cities is evidenced by the region's disproportionate share of the nation's largest hospitals. Although southern states contain approximately 21 percent of the nation's population, medium and large urban centers in the South contain more than 40 percent of the nation's hospitals with 500 or more beds. Furthermore, the ratio of hospital beds to the population in the largest southern urban areas is more than 10 percent higher than it is for the United States as a whole, indicating the inordinate significance of health care in southern cities. The importance of health care in these urban centers is reinforced in that not only are there more hospitals but they are intensively utilized. The hospitalization rate for the major southern urban centers is 25 percent higher than the national average, reflecting the influence of physicians and hospitals on their respective communities.

RICHARD K. THOMAS
University of Tennessee Health Science Center

American Hospital Association, *AHA Guide* (2009); U.S. Census Bureau, *Census of Population* (2000).

Medical Science, Racial Ideology, and Practice, to Reconstruction

Slavery in the American South presented a unique catalyst for growing questions and concerns about race within the practice of medicine. The arrival of the first Africans in Jamestown, Va., in 1619 laid the foundation for future medical preoccupations with physical differences between blacks and whites. Ideas governing disease causation remained closely tied to the relationship between region, climate, and physical constitution and the South's distinctive ecology—for example, hot and humid summers and frequent spells of yellow fever led to pervasive beliefs that the southern climate was too harsh for the constitutions of white laborers but perfectly suited for Africans. Southern physicians made broad assessments about morbidity and mortality along racial lines; most notable was the enduring belief that African Americans were immune to yellow and malarial fevers.

Whites saw black skin as both a biological advantage and a mark of degeneracy because, while it offered protection from the ill effects of the climate, it was also viewed as a sign of primitiveness. In *Notes on the State of Virginia* (1787), Thomas Jefferson lamented "Negroes" were too different from whites for equal coexistence because their physical differences were "fixed in nature." In the 1850s Josiah C. Nott, a respected and outspoken southern physician, cited black features as evidence of separate origins of the races of mankind—a controversial theory known as polygenesis. Polygenesis remained the hallmark of Nott's race theories and the cornerstone of his beliefs in black biological inferiority.

Even as white physicians claimed African Americans were biologically different from whites, they relied on African American bodies for dissections, experiments, and anatomical demonstrations. In Charleston, S.C., the medical college made no secret of its use of "anatomical material" taken from the "colored population." Southern physician J. Marion Sims, a pioneer in surgical techniques and "father of gynecology," used three female slaves, Lucy, Betsy, and Anarcha, to find a treatment for vesicovaginal fistulas. Despite his success, and the relief his new treatment offered to women of all races, his rise to fame rested upon experimental surgeries performed on enslaved African American women.

Bold claims about black inferiority gained momentum in the decades leading up to the Civil War. Proslavery medical thought worked in tandem with "*Herrenvolk* democracy" to justify attitudes of white supremacy. Southern physicians fashioned themselves as experts on slave health, and antebellum medical articles on "distinctive Negro" physiology proliferated in medical journals. Samuel A. Cartwright, onetime chairman of the Medical Association of Louisiana, authored "A Report on the Diseases and Peculiarities of the Negro Race," published in 1851 in the *New Orleans Medical and Surgical Journal* and later reprinted in *De Bow's Review*. Cartwright's report was more of a political invective interspersed with spurious medical claims than a medical overview of African American diseases, and in it he infamously coined the term "drapetomania"—the disease causing slaves to run away.

The Reconstruction era engendered new trends in medical ideas about race, as well as views of slavery, that were more nostalgic than accurate. Whites argued that, before the Civil War, slaves' health was guaranteed by their benevolent masters, and the troubles that freed African Americans faced after Emancipation were consequences of their inability to look after their own welfare. Physicians pathologized freedom and published articles on increases in the rates of tuberculosis and insanity among African Americans in the South. At the Alabama Insane Hospital, freedom was cited as a cause of insanity among African American patients. The traits of African Americans, before seen as advantages for laboring in the South, became the loci of their postbellum health complaints. Social Darwinists claimed African Americans would become extinct because they could not compete in society at the same level as whites. In the aftermath of slavery, African Americans continued to face challenges in gaining control over their health care as a result of deliberate and institutionalized neglect caused by Jim Crow policies at the turn of the century.

RANA ASALI HOGARTH
Yale University

George M. Frederickson, *The Black Image in the White Mind: The Debate on Afro-American Character and Destiny, 1817–1914* (1987); John S. Haller Jr., *Medical History* (July 1972); Reginald Horsman, *Josiah Nott of Mobile: Southerner, Physician, and Racial Theorist* (1987); John S. Hughes, *Journal of Southern History* (August 1992); Kenneth Merrill Lynch, *Medical Schooling in South Carolina, 1823–1969* (1970); Gary Puckrein, *Journal of American Studies* (August 1979); Todd L. Savitt, *Journal of Southern History* (August 1982), *Medicine and Slavery: The Disease and Health Care of Blacks in Antebellum Virginia* (1978); Alden T. Vaughan, *William and Mary Quarterly* (July 1972); John Harley Warner, in *Science and Medicine in the Old South*, ed. Ronald L. Numbers and Todd L. Savitt (1989); Christian Warren, *Journal of Southern History* (February 1997).

Medicine, States' Rights

As slavery came under increasing attack in the 30 years before the Civil War, white southerners in all fields closed ranks. The professions, as might be expected, sought to provide the intellectual justification for the South's peculiar institution—lawyers argued on constitutional grounds, ministers cited the Bible, and physicians, as natural scientists, endeavored to demonstrate that blacks were an inferior race and that southern medicine was distinct from northern medicine. The first southern physician to argue the inferiority of the black race was Dr. Josiah C. Nott, a prominent physician in Mobile, Ala., who had learned his medicine in the best northern and European medical schools. In the early 1840s he published an article maintaining that blacks had less endurance, had shorter lives, and were less prolific than whites, and that intermarriage could only result in the destruction of both races. By 1850 he was arguing that "an immutable law of nature" made it impossible for blacks to become civilized. Following his argument to its logical conclusion, he asserted that blacks were "better off in slavery [in] the South than in freedom elsewhere."

Because Nott argued that blacks were an entirely separate race from whites, his views, by conflicting with the biblical account that man descended from Adam and Eve, ran counter to the religious fundamentalism of the South. Much more to the liking of white southerners in general were the ideas of the physiological school, represented by physicians who accepted the biblical view of creation but still maintained that striking anatomical and physiological differences separated whites and blacks. The leading exponent of this school was Dr. Samuel A. Cartwright of Natchez, Miss., and New Orleans. Like Nott, Cartwright was well trained and was widely recognized for his scientific work. He was also a born controversialist who loved to express his views. Beginning in the 1840s he wrote a stream of articles on blacks and southern medicine. In 1850 he was appointed chairman of a committee of the Medical Association of Louisiana to investigate the diseases and physical peculiarities of blacks. His report declared that the "shade of pervading darkness" was present not only in the skin but throughout all parts of the body—even including the "fluids and secretions." The skeletal structure of blacks was distinct from whites, the brain 10 or 11 percent smaller, the vascular system much less developed, and the lungs smaller. Moreover, their brains were so constituted to produce an excess of "nervous matter," which would have made blacks unmanageable had it not been for the "deficiency of red blood" owing to the inefficiency of their lungs.

Cartwright's studies also revealed two new diseases among slaves—one was "drapetomania," a disorder that caused them to run away, and the other was "dysaethesia aethiopica," generally known to overseers as "rascality." As his studies continued, Cartwright was led inexorably to the conclusion that the fundamental differences between the two races meant that "the same medical treatment which would benefit or cure a white man, would often injure or kill a Negro." The ideas of Nott, Cartwright, and other medical writers confirmed what many southern physicians had been thinking, and in the ensuing years a flood of articles in southern medical journals cited more and more "scientific" evidence of black inferiority.

Cartwright's argument for a distinctive form of southern medical practice also had a receptive audience. In the long quest for causes of disease, men had constantly studied the role of climate, meteorology, and topography; white southerners believed these factors affected their health. Cartwright insisted not only that the diseases of blacks and their treatment were distinct from those of whites but that southern diseases in general were different from those in the North. Like the anatomical argument, the thesis about southern diseases contained a modicum of truth. Yellow fever was essentially a southern problem at that time, malaria and enteric disorders did tend to be more acute in the South than in the North, and southern blacks did have a greater degree of immunity to certain disorders.

If southern medicine was distinct, it was logical that southern practitioners should be trained in the South. Although American medicine, with some exceptions, generally lagged behind that of western Europe, medical education in the South was not even up to the level provided by the better northern schools; consequently, southern medical students seeking the best training went first to the North and then to Europe. The South in the antebellum years was trying to gain economic independence from the North, and keeping its medical students at home was financially advantageous. Southern businessmen were acutely conscious of the considerable economic loss entailed by training over half of all southern medical students in the North.

Added to this was the self-interest of southern medical schools. Nearly all medical schools in the United States were proprietary institutions dependent upon student fees for revenue; hence, there was keen competition for students. Southern medical journals and newspapers beginning in the early 1850s constantly emphasized the distinctiveness of southern medicine and the need for southern practitioners to be trained in the South. As the idea of separate southern medicine gained credence, enrollment in southern schools steadily increased, and new medical schools appeared on the scene. The campaign to keep southern medical students at home culminated in the mass resignation in the 1859–60 academic year of some 200 students from Jefferson Medical College in Philadelphia and another 100 from the University of Pennsylvania.

By 1860 only a few skeptics in the South questioned the concept of a specific southern medicine; the vast majority of southern physicians never doubted that the diseases they treated and the methods they used differed from those in the North.

JOHN DUFFY
University of Maryland

James O. Breeden, *Bulletin of the New York Academy of Medicine* (1976); John Duffy, *Journal of Southern History* (May 1968), ed., *The Rudolph Matas History of Medicine*

(1962); Harold Jackson, in *Bones in the Basement: Postmortem Racism in Nineteenth-Century Medical Training*, ed. Robert L. Blakely and Judith M. Harrington (1997); Mary Louise Marshall, *New Orleans Medical and Surgical Journal* (1940–41); Todd L. Savitt, *Race and Medicine in Nineteenth- and Early Twentieth-Century America* (2007).

Obesity

In the past two decades, waistlines across the county have expanded, but in the South the increase in obesity rates can best be described as an epidemic. An estimated 80 percent of counties in the South and Appalachia have elevated rates of obesity and diabetes. A study published jointly in 2010 by the Trust for America's Health and the Robert Johnson Wood Foundation reported that 9 of the 10 states with the highest rates of adult and childhood obesity are located in the South. Additionally, the South leads the nation in percentages of hypertension and diabetes. Some attribute high obesity rates to a regional culture that embraces the consumption of high-fat foods as heritage, but socioeconomic factors, environment, and food security are issues as influential to the well-being of southerners today as they were more than 100 years ago. It is still primarily the low-income residents of remote, rural areas who are struggling for access to fresh foods, but instead of the cornbread and fatback that was popular during the early 20th century, southerners today have an array of quick, processed meals at their disposal.

Despite the stereotypes of southerners as fat perpetuated by television characters like Boss Hogg from the *Dukes of Hazzard*, southerners have not always been overweight. White and black sharecroppers in the rural South lived in abject poverty. Those with the means to do so planted small subsistence gardens, but most survived on purchased cornmeal and fatback. Pellagra, the result of niacin deficiency, and malnutrition ran rampant among the rural population of America. In times of plenty, southerners ate biscuits, pork, cowpeas, and sweet potatoes. Despite a diet high in fat and carbohydrates, southerners were no heavier than the average American. In 1960, 13 percent of Americans were obese. As obesity rates rose across the nation, those in the South and Mid-Atlantic seem to grow at almost the same rate. By 1999, 26 percent of Americans were obese, and southern states had the highest fattest populations.

The South is home to 9 of the 10 states with the highest poverty rates in the nation, and the statistical correlation between poverty and obesity is not coincidental. Poverty and environment work simultaneously to influence obesity rates. Though the region is evolving rapidly, its environment is still primarily rural. It is not uncommon for residents to live more than three miles from a grocery store, and public transportation and pedestrian walkways are scarce. Lack of money and isolation qualify many rural, low-income regions throughout the South as "food deserts": geographic areas with limited access to conventional grocery stores. In these food insecure neighborhoods, residents subsist off of sundries from local corner stores or fast-food restaurants. Fresh fruits and vegetables stocked in these stores come at a premium, forcing low-income residents to choose between high-priced produce and low-cost, shelf-stable products. Cheaper still is the value menu from a fast-food outlet, where two dollars can purchase a complete dinner.

Childhood obesity has become cause for alarm in the South as well. Nine of the 10 states with the highest rates of childhood obesity and diabetes are located in the South. If childhood obesity rates do not decrease, the next generation of children may have a life-span shorter than their parents'. Schools have become instrumental in combating childhood obesity, in conjunction with the federal "Let's Move" campaign championed by First Lady Michelle Obama. Through the Farm to School program, local farms supply a portion of schools' food. School gardens that reinforce students' agricultural heritage are also reminiscent of the subsistence garden that families relied on before the commercialization of food processing. Church and community gardens have functioned similarly in combating adult obesity.

The South's struggle with obesity has been well documented in the media. Mississippi was scrutinized in 2010 when Trust for America's Health and Robert Johnson Wood Foundation ranked it the fattest state in America for the sixth year running, leading the nation in obesity, hypertension, and diabetes rates. In 2008 three members of the state House of Representatives issued a bill that would prohibit state-licensed restaurants from serving obese patrons. Though officials claimed the bill was meant to draw attention to the severity of Mississippi's obesity problem, the media condemned its insensitivity to overweight residents. Alabama, site of the third-highest obesity rate in the nation, caused intense debate after it proposed a "fat tax" on state employees. Obese state employees who did not attempt to lose weight would be charged a $25 surcharge on their health insurance. In the summer of 2010, the ABC television network partnered with celebrity chef Jamie Oliver for an American rendition of his television show *Jamie Oliver's Food Revolution*. Oliver traveled to Huntington, W.Va.—the fattest city in America—to educate its citizens about the importance of nutrition and exercise. Although the show was not widely popular, it was awarded an Emmy for Outstanding Reality Program.

With increased publicity about the dangers of obesity, southerners are making changes to promote healthful lifestyles. At the state level, Tennessee is considering legislation that would give low-interest loans for the development of grocery stores in underserved neighborhoods. Louisiana has approved the creation of a food policy council that will find and develop opportunities for sustainable food programs in the state. Nearly three-quarters of Mississippi high schools have banned the sale of soft drinks on campus.

Though obesity remains a prevalent issue in the South, community leaders are taking action to address the underlying causes of obesity: food insecurity, lack of physical activity, and poverty.

NOVELETTE BROWN
University of Mississippi

Nanci Hellmich, *USA Today* (3 August 2010); Kenneth M. Johnson, in *Challenges for Rural America in the Twenty-first Century*, ed. David L. Brown and Louis E. Swanson (2003); Harvey A. Levenstein, *Paradox of Plenty: A Social History of Eating in Modern America* (2003); Michel Martin, *Mississippi Town Joins Plan to End Child Obesity*, National Public Radio (aired 15 February 2010); Lois Wright Morton and Troy C. Blanchard, *Rural Realities* 1:4 (2007); Jamie Oliver, *Jamie Oliver's Food Revolution* (2010, television); Claire Suddath, *Time* (9 July 2009); U.S. Department of Agriculture, *Food Environment Atlas*; Michelle Van Ploeg et al., *Access to Affordable and Nutritious Food—Measuring and Understanding Food Deserts and Their Consequences: Report to Congress* (2009).

Physicians, African American

After Emancipation, a small but important black medical community developed in the South as medical schools opened to educate African Americans. However, as a result of medical reforms, in the first three decades of the 20th century medical schools for African Americans in New Orleans, Raleigh, Chattanooga, Louisville, and Memphis all closed their doors, leaving Howard University (Washington, D.C.) and Meharry Medical College (Nashville, Tenn.) as the only institutions south of the Mason-Dixon line where African Americans could receive a medical education. Although a handful of medical schools outside the region annually admitted a small number of black students, the result of the closing of these schools was that the number of practicing black physicians in the nation remained stagnant for more than 20 years, at just under 4,000. No state medical school in the South admitted black students until 1948, when Edith Mae Irby was admitted to the University of Arkansas Medical School. By 1957, 14 of the South's 25 formerly all-white medical schools admitted African Americans, but it would not be until 1969 that a majority of the nation's black physicians graduated from schools other than Howard and Meharry.

Postgraduate education was an even more difficult hurdle for African American physicians to overcome. As medical education developed, intern and residency training became a standard part of a physician's education in the 20th century. Until the 1930s, however, there were not enough internships available for the number of African Americans graduating from medical school each year, a situation that was tied directly to race, as many hospitals chose to leave internships unfilled rather than accept an African American. This was not specifically a southern problem either, as most hospitals outside the South refused to accept black interns, because there was an aversion to having black doctors treat white patients, especially women. There were even some all-black facilities, such at Atlanta's Grady Hospital, which hired only white interns, as many white administrators saw all-black hospitals as the perfect venue for training white physicians. To try to fill the void, a small number of black hospitals, such as St. Louis's Homer G. Phillips and Washington, D.C.'s Freedmen's Hospital, provided more internships than they needed, and numerous small, ill-equipped black hospitals in the South offered unaccredited internships.

If obtaining an internship was difficult for a black physician, obtaining a residency anywhere in the United States before the 1940s was virtually impossible. In 1930 there were no residencies available for African Americans in any specialty field; black hospitals simply did not have the facilities and physicians to train specialists, and white hospitals refused to admit black doctors to their programs. As a result, there were fewer than 25 black specialists practicing in the United States in 1939. The inability of African Americans to secure internship and residency training not only hampered their professional development but also contributed to the perception that black physicians were not as well trained as their white counterparts, because, in fact, they often were not.

Once African American physicians received their education and decided to settle in the South, they faced unwanted—and often unexpected—competition for patients from white physicians. Although caste usually dictated that black physicians not treat whites, white doctors were often more than willing to treat African Americans who could pay, even while relegating them to discourtesies such as segregated waiting rooms. Black physicians tried to exploit this lack of respect as a means to lure black patients into their own offices, but African Americans had a host of legitimate reasons for choosing white physicians, despite the humiliations of Jim Crow care. Because of either discrimination or expense, black doctors often could not provide patients with the same services as

Unidentified physician with child, Kentucky, 1950s
(Photographic Archives, University of Louisville, Kentucky)

their white counterparts, as white doctors usually had better-equipped offices with more modern facilities and medicines than did blacks. Many African Americans also went to the doctor their employer told them to go to, and white employers often steered their employees to white physicians—especially if the employer was paying the bill. White physicians had an advantage in attracting patients, recalled Mississippi's B. L. Bell in 1939, because they "can work through the people that the Negroes work for."

Once established, the greatest handicap black physicians faced was a lack of hospital access. African Americans were barred from many hospitals in the South during the Jim Crow era, and most hospitals that admitted black patients usually did so only in segregated basement wards. The shortage of hospital beds was not the only problem for black physicians. Many hospitals that admitted African Americans often did not grant black physicians staff privileges, even if the patient requested that they be treated by their personal physician. African American physicians were therefore forced to turn many of their patients over to white doctors for hospital care. Unable to treat patients in their local hospitals, they lost potential income from surgery and other hospital treatments. Income was also forfeited, as white doctors were usually more than willing to poach the general practice of a patient whom a black doctor turned over for hospital treatment. As Dr. Albert Dumas of Natchez, Miss., lamented in 1910: "[White surgeons] carry the patient to the hospital and often you never see them again."

Excluded from large, public, well-equipped hospitals, African American physicians often founded small, private hospitals and clinics, such as the Burruss Sanitarium in Augusta, Ga. By the 1930s more than 200 small, independent black-run hospitals had opened in the United States. Serving an impoverished clientele, physicians constantly struggled to keep their small hospitals open. African American organizations in some southern communities also built small hospitals to fill the need for black hospital beds. Many of these hospitals were owned and run by fraternal organizations, such as the Afro-American Sons and Daughter's Hospital in Yazoo City, Miss. Other black-run hospitals were administered by black colleges and medical schools, some of which, like New Orleans's Flint-Goodridge Hospital, survived even after the medical school folded.

Realizing that hospital care was often an expense black southerners could not afford, many physicians gave their time to bring health care to African American communities at little or no cost. Health clinics were regularly sponsored by black hospitals, medical schools, churches, or philanthropic organizations in the Jim Crow South, where physicians donated their time and services. The most famous health clinic for black southerners occurred annually at Tuskegee's John Andrew Hospital from 1912 to 1969. The all-black National Medical Association (NMA) held its annual meeting at Tuskegee Institute in 1912, and, in conjunction with the convention, a clinic was held at the school's John A. Andrew Hospital, organized by Tuskegee Institute's Dr. John Kenney. At the initial clinic, NMA physicians treated more than 400 persons and performed 36 operations. The success of the clinic convinced Kenney to make the service an annual event, and physicians and surgeons from across the nation came to Tuskegee each spring to donate their services.

The National Medical Association was founded in 1895 as a counterpart to the segregated American Medical Association (AMA). Like the AMA, the NMA drew its support from state and local affiliates. Black doctors formed more than 100 local medical societies throughout the South during the segregation era to share professional information, hold clinics, promote medical advancements, and provide postgraduate training for their members. Society meetings often included the presentation of scientific papers, discussions of new advancements in the medical field, and lectures on modern health problems. These meetings also provided badly needed professional companionship for black doctors who were frequently ostracized by white physicians in their towns. As Gilbert Mason, secretary of the Gulf Coast Medical, Dental, and Pharmaceutical Association (of Mississippi, Alabama, and Florida) recalled, "We saw to it that black practitioners did not work in total isolation." Societies also sometimes worked with white organizations, hospitals, and schools to improve the quality of black doctors. White physicians regularly gave lectures and demonstrations at black medical society meetings, even in the Deep South. Through their societies, African American physicians sometimes even gained access to segregated facilities that were normally off-limits to them. For example, during the 1930s members of North Carolina's Old North State Medical Society regularly attended postgraduate seminars and visited pathological clinics at Duke University Hospital, a facility from which they were otherwise barred.

However, as a result of discrimination, violence, lack of opportunities, and the migration of African Americans from the region, black physicians steadily left the South as the 20th century progressed. By 1930, 40 percent of the nation's 3,805 black physicians resided outside of the South, where 80 percent of the nation's almost 12 million African Americans still lived, and the majority of those who remained were located in urban areas. In rural states, the situation was most acute. Mississippi saw its number of black physicians decline from 71 in 1930 to 55 in 1940, and 56 of the state's 82 counties had no black physicians at all. Even southern cities saw a decline of black physicians as the century progressed. New Orleans, once a hub of the black medical community, saw the number of black physicians practicing in the city decline from 50 in the 1930s to only 16 by the mid-1950s.

To combat discrimination and stem the loss of black physicians from the region, both the NMA and the NAACP fought to end segregated medical facilities. In 1963 the federal bench handed down the landmark case in the fight for hospital integration in *Simkins v. Moses H. Cone Memorial Hospital*. It took the enforcement of the *Simkins* case, along with the Civil Rights Act of 1964 and the enactment of Medicare (which withheld federal funding from segregated facilities), to finally bring integration to the South's hospitals. Between 1964 and 1970, most southern hospitals accepted the integration of both black patients and staff, and AMA-affiliated societies also began to drop their racial bars. As the remaining vestiges of medical segregation died away, African American physicians were finally able to move into the mainstream of the southern medical establishment.

THOMAS J. WARD JR.
Spring Hill College

Edward Beardsley, *A History of Neglect: Health Care for Blacks and Mill Workers in the Twentieth-Century South* (1987); Herbert M. Morais, *The History of the Negro in Medicine* (1967, 1976); Todd L. Savitt, *Bulletin of the History of Medicine* (Winter 1987); Thomas J. Ward Jr., *Black Physicians in the Jim Crow South* (2003).

Poverty, Effects of

Beneath mythical images of parasol-accessorized southern belles sashaying amid goateed gentlemen in white linen suites sitting under the shade of sprawling trees festooned with gently swaying curtains of Spanish moss lurks the reality of poverty in the South. Poverty has unfortunately been a defining feature of the South, Old and New. In a classic essay, Avery Craven observed that "no generalization can hold from place to place and from time to time [in the South]; that its life was so individualized by rural and frontier forces that each relationship was a thing unique in itself; that the human element ever looms large in rural worlds and the accidents of health, weather, and personal qualities rise in proportion." Perhaps, but Craven also noted the common ties that bind southern blacks and whites together in a struggle against disease, namely, their "strikingly similar" diets and "homespun" dress that seldom included shoes; in addition, their "near-nakedness furnished subject to the traveler for comment with about equal frequency."

Those two features—diet and dress (along with one other to be discussed later)—would define the disease experience of whites and blacks living in poverty throughout the South. When the devastation of war combined with the boll weevil to shift the cotton-based economy of the region toward a "New South" built upon northern industrialists' eagerness for a cheap and ready labor supply, there is some indication that health conditions worsened appreciably for the poor. Thousands gave up farming and took up subsistence wage jobs in mines and textile factories, leaving their vegetable patches behind. Similarly, poor tenant farmers who stayed tied to the land found themselves pushed for maximum crop yields by profit-seeking landholders who saw small vegetable gardens that normally supplemented the family table as a waste of otherwise valuable cotton-producing soil. Either way, both suffered greatly by a diet limited to the company store or seasonal incomes that left the body subject to pellagra, a nutritional deficiency disease (specifically, a deficiency of niacin) that became epidemic throughout the South in the early years of the 20th century. From 1907 to 1911, for example, eight southern states reported more than 15,000 cases of this debilitating ailment that often started with a red dermatitis, progressed to diarrhea, and in severe cases led to dementia and even death. That same group of states witnessed an astonishing 39 percent mortality rate from this wholly preventable disease. Joseph Goldberger, the U.S. Public Health Service official who correctly identified pellagra as a nutritional deficiency despite colleagues who insisted it was an infectious disorder, rightly commented that "the problem of pellagra is in the main a problem of poverty. Education of the people will help; but improvement in basic economic conditions alone can be expected to heal this festering ulcer in the body of our people." Not until World War II demanded a fit and ready fighting force did the nation solve the pellagra problem with a mandated food enrichment program.

Lack of shoes (especially among children) caused a similar pandemic of hookworm. Although this disease was certainly a serious problem well before the 20th century and likely endemic among the poor of the antebellum South, it was Charles Stiles's shocking exposé in 1902 that brought hookworm to national attention. Here bare feet and lack of properly drained privies combined to provide a perfect environment for fecal-borne larvae to infect tens of thousands of the southern poor—in 1905 as much as 40 percent. This scourge prompted the Rockefeller Foundation in 1909 to launch a commission to fight hookworm in the South. Progress was slow. A 1914 survey of southern schoolchildren showed an infection rate of 39 percent. Fortunately, concerted public health efforts at state, local, and national levels over the next half century would make a dramatic impact in reducing hookworm infection rates.

If diet and dress (or lack thereof) affected the southern poor, so too did climate. The subtropical conditions bred an abundant supply of the *Anopheles* mosquito responsible for malaria—a "fever" common to swampy regions of the South from colonial times. Modern advances in detection of infection rates and identification of the organism responsible for the disease allowed for precision in determining the extent of this southern debilitator. Plasmodium smears were positive among poor whites in 5 to 13 percent of the cases with positive infection rates more than twice that among black sharecroppers who spent much of their time in fields swarming with mosquitos. The mid-1930s still witnessed more than 135,000 cases of malaria with a mortality rate in excess of 4,000. As in the case with hookworm, ambitious public sanitation campaigns helped eradicate the poorly drained sewage systems and pools of stagnant water in which the mosquito laid its eggs. By 1950 malaria was largely a thing of the past in the American South.

These three diseases—pellagra, hookworm, and malaria—interestingly have marked symptomatic commonalities and each played a part in stigmatizing the South. All tend to be chronic ailments that tax the system and lead to secondary vitamin deficiencies and other health problems. These maladies also commonly manifest themselves in mental acuity and cognition deficits. Long recognized in cases of pellagra and hookworm, the impact of malaria upon the mental attributes of its suffers was also noted by the Rockefeller Commission for the Study and Prevention of Malaria in 1913. The commission reported that malaria "leaves its subjects anemic and neurotic and is responsible for inertia, loss of will power, intemperance, and general mental and moral degradation." The notion of the so-called lazy southerner was largely a product of these three diseases intrinsic to the poverty commonplace in the South, and each was driven by dietary habits, climate, custom, and socioeconomic conditions in some measure unique to the region. Interestingly, the vitamin deficiencies incident with hookworm can prompt its victims to geophagia (soil eating). It may be that the often derisive reference to southern "clay eaters" is actually a manifestation of disease common to the region. Fortunately, the dimwitted "country rube" of the South vanished with the diseases that lent themselves to this cartoonish stereotype. Although surely many ailments afflicted southerners as a result of conditions of dietary deprivation and substandard living conditions, none was more powerful in defining—indeed, in caricaturing—the region than pellagra, hookworm, and malaria.

It would be wrong to assume, however, that the effects of poverty on the health and well-being of southerners ended in the mid-20th century. Amid the War on Poverty in the 1960s, six physicians sent by the Field Foundation to examine health conditions among the rural poor of the Mississippi Delta were appalled by conditions similar to those they had seen in developing countries in Africa and Asia. Calling the conditions "shocking," the group "found it hard to believe we were examining American children of the 20th century."

Inordinately high rates of disease in the South are still a result of poverty levels that remain well above the 14.2 percent national average. Mississippi, for example, leads the nation with a poverty rate of 21.3 percent, with Louisiana's and Georgia's 18 and 17.5 respective levels not far behind. Although it has been assumed that hookworm has been eradicated, Peter Hotez points out that no large study has been published recently and "studies conducted during the late 1970s and 1980s revealed that significant numbers of American schoolchildren living in poor areas of the American South were infected with the large common roundworm, *Ascaris lumbricoides*, including 32 percent of school-aged children living in an unincorporated area of northern Florida." Other insidious killers such as diabetes, high blood pressure, and now AIDS find a higher incidence among the southern poor today than the national average.

Although poverty and disease have dogged the South, efforts to combat both continue. Until state budgets, income levels, and public health initiatives witness substantial improvements, the synergy between poor living conditions and poor health will undoubtedly continue. Exacerbated by inadequate access to health care—an especially acute problem in the rural South—the untoward effects of poverty on the health of the South is likely to persist.

MICHAEL A. FLANNERY
University of Alabama at Birmingham

William M. Brewer, *Journal of Negro History* (January 1930); Avery O. Craven, *Journal of Negro History* (January 1930); John Ettling, *The Germ of Laziness: Rockefeller Philanthropy and Public Health in the New South* (1981); Christine Gorman, *Scientific American* (July 2011); Peter J. Hotez, *PLoS Neglected Tropical Diseases*, www.plosntds.org (26 December 2007); Institute for Research on Poverty, University of Wisconsin–Madison, www.irp.wisc.edu; Greta Jong, *Journal of African American History* (Autumn 2005); Stephen J. Kunitz, *Journal of Health and Social Behavior* (June 1988); Mike G. Martin and Margaret E. Humphreys, *Southern Medical Journal* (August 2006).

Professionalization of Science

Although hardly an ordinary man, Andrew Jackson, with his election to the White House in 1828, signaled the "Age of the Common Man" in America. Hard work, determination, and self-improvement were the catchwords of the day, and elitism was out of fashion. Public insistence that knowledge be immediately understandable and useful aroused concern among a growing number of scientists. Scientists resented their need to supplement their meager incomes, usually derived from college professorships, by traveling the popular lecture circuit to which Americans flocked. The root of their discomfort lay in their awareness that rapidly expanding scientific knowledge, and the means by which this knowledge was acquired, could no longer be communicated to a general audience.

As scientific inquiry, fueled by the growth of experimentation, expanded knowledge beyond the realm of the layman's understanding, fields that once belonged in the public domain came to be dominated by small, select groups of scholars with specialized educations. Pleading the necessity of basic research, these scientists maintained that their efforts to advance man's knowledge of his world could proceed only as the result of careful observation and experimentation, free of the financier's watchful eye and the public's preference for utility. These men gained a growing respect for one another as professionals and were anxious that the general public afford them the same professional status as that enjoyed by physicians, lawyers, and clergymen. They struggled to disassociate themselves from technical inventors and, more importantly, from quacks and charlatans who touted miracle cures and regaled the public with outrageous "corrections" of accepted scientific principles.

To foster a spirit of professionalism, American scientists needed contact among themselves. The American Association for the Advancement of Science (AAAS) was founded in 1848 to encourage scientific research, arbitrate scientific disputes, and weed out "pretenders" to the profession. By the end of the century, more specialized societies emerged, including the American Chemical Society, the Geological Society of America, and the American Mathematical Society. All of these organizations defined their profession to include a specialized graduate degree, employment as a scientist, and evidence of scholarly research.

Most antebellum southern scientists remained on the periphery of this movement. Although some of them maintained membership in the AAAS, they seldom attended national meetings, usually held in the northeastern region of the nation, because of the distance involved, inadequate transportation, and financial pressures. However, the long-perpetuated notion of the antebellum South as "essentially unscientific" simply is not true. Planters such as Henry William Ravenel, James Hamilton Couper, and Benjamin L. C. Wailes maintained an active correspondence with scientists throughout the world and enjoyed enviable reputations for the specimens, drawings, and descriptions of southern flora and fauna that they provided.

Scientists in the emerging professional sense also inhabited the cotton kingdom. Thomas Cooper, John and Joseph LeConte, Frederick A. P. Barnard, and Elisha Mitchell, all college professors, earned the respect of their colleagues nationwide. Other professionals included Denison Olmsted and Michael Tuomey, who headed the state-sponsored geological surveys of North Carolina and Alabama. Charleston, S.C., the region's major metropolitan area, boasted the greatest concentration of scientific talent. With its relatively large, stable, and wealthy population and intellectual centers such as the Charleston Museum and the College of Charleston, the city sustained such notables as John Bachman, Lewis R. Gibbes, John McCrady, Edmund Ravenel, and Francis S. Holmes.

The move toward professionalization in the South, hardly begun by 1860, suffered in the years after the Civil War. Colleges throughout the region, the primary employers of scientists, closed their doors for a varying number of years during Reconstruction. Some, such as the University of Alabama, lay in smoldering ruins. Libraries and laboratories, ravaged by neglect as much as by war, had to be rebuilt from scratch. Faculty members were scattered. Some chose alternative careers; others, such as John and Joseph LeConte, grew discouraged with Reconstruction politics and moved elsewhere. Young men seeking a science education attended northern or German universities, and many of those who thus imbibed the spirit of research sought positions in schools that would support such work. Southern institutions of higher education, slowly re-opening during the 1870s, could scarcely scrape together sufficient funds to pay their faculty members; they provided little in the way of compensated time or financial incentive for research.

Nonetheless, a few well-trained men of science did serve in the postbellum South's colleges, among them Eugene Allen Smith of the University of Alabama, Francis Preston Venable of the University of North Carolina, and William Louis Jones of the University of Georgia. Although pleased that southern colleges were reopening and even expanding their curricula, these men, all products of a graduate education that emphasized the spirit of scientific inquiry through research, suffered from isolation, restricted budgets, and poor communication and transportation facilities. Hoping to "ward off the deadening effect that isolation was bound to have upon our scientific work," Venable and three of his colleagues at the University of North Carolina in 1883 organized the Elisha Mitchell Scientific Society. It met monthly during the academic year to provide North Carolina scientists contact with one another and an opportunity to share research efforts; in addition the society published quarterly the *Journal of the Elisha Mitchell Scientific Society*. However, neither the Mitchell Society nor a similar group in Alabama, the Alabama Industrial and Scientific Society, was able to serve the needs of the relatively few and scattered southern scientists of the late 19th century. Attracting little popular support and receiving almost no institutional financial aid, the societies could not continue their activities. The Alabama organization collapsed completely, and the Mitchell Society survived only as a local university forum.

The pool of southern scientists continued to increase, however, with the growth of southern colleges and universities under the leadership of progressive educators who witnessed the expansion of northern and midwestern universities and coveted for their own region similar educational advantages. The best indicator of the professionalism of this growing body of pre–World War II southern scientists lies once again in their organizations. Still isolated from colleagues in other regions of the nation, they formed state academies of science to provide contact with one another, to foster the spirit of research, and to offer at least a modest outlet for publication. By the mid-1930s, every southern state boasted a state academy of science; by 1940 approximately 3,500 scientists throughout the South, over 90 percent of whom were college and university faculty members, belonged to these academies. An ever-increasing number of these persons held the doctoral degree, and, if only a handful of them were on the cutting edge of scientific research, many others remained informed and conducted research of local interest and benefit.

Before World War II, very few scientists were employed by industry or independent research laboratories. Most southern industry was either extractive, as in the case of the copper mines of Tennessee and the iron-ore mining operations of north Alabama, or labor intensive, as with the tobacco factories, textile mills, and fertilizer plants that dotted the landscape. With profits more dependent on a labor force willing to work for low wages than on increased efficiency through research and development, management saw no reason to employ professional scientists. Research laboratories such as those at United States Steel and Bell Telephone, an integral part of northern industry by World War I, would not appear in the South for another generation.

World War II and the accompanying financial boom changed the South drastically and permanently. Rapid industrial development brought research-oriented corporations into the region as southern states competed for their attention with such incentives as low taxes and prime locations. Independent research laboratories emerged as well. Areas such as the Research Triangle Park near Raleigh, N.C., now rival any other region of the nation for productive, scholarly, scientific research. Universities blossomed, too, thanks to greatly increased financial support. Graduate schools attracted highly qualified professors; increasingly, southerners not only chose to remain at "home" for their education but often accepted permanent employment in the region. Modern transportation and communication, coupled with institutional funding for travel and research, meant that these scientists were no longer isolated as they had been earlier. By the 1970s and 1980s, black and women scientists represented a growing, previously excluded human resource for the South. Southern scientists have earned the respect of their colleagues, both personally and for the institutions that now support their endeavors.

Research universities in the South at the start of the 21st century stand literally and symbolically as "steeples of excellence" in the region's cultural and institutional landscape. Foremost are the 11 southern universities among the 59 universities nationwide selected for membership in the prestigious Association of American Universities (AAU), which "focuses on national and institutional issues that are important to research-intensive universities, including funding for research, research and education policy, and graduate and undergraduate education." The 11 distinguished universities are, with parenthetical reference to year of selection, as follows: the University of Virginia (1904), the University of North Carolina at Chapel Hill (1922), the University of Texas at Austin (1929), Duke University (1938), Vanderbilt University (1950), Tulane University (1958), Rice University (1985), University of Florida (1985), Emory University (1995), Texas A&M University (2001), and the Georgia Institute of Technology (2010).

The 11 AAU member institutions are then joined by other universities in the South whose records in obtaining sponsored research grants along with conferring Ph.D. degrees are recognized as major research institutions. This tier includes Louisiana State University at Baton Rouge, the University of Georgia, the University of Kentucky, North Carolina State University, Virginia Tech University, the University of Tennessee, University of South Florida, University of Miami, Clemson University, University of Oklahoma, Auburn University, the University of Mississippi, the University of Alabama, Mississippi State University, the University of Arkansas, and the University of South Carolina.

This group of 27 universities represents 13 states, suggesting a reasonable distribution of scholarly talent across the region. Most remarkable is the chronology of membership in the Association of American Universities. No university in the South was invited to be a charter member of the Association of American Universities in 1900. Only five southern universities were members at the end of World War II. One signal of the sustained ascent of research universities in the South is that four were selected for AAU membership since 1985, illustrative of an acceleration of academic research in the region in the late 20th century. In sum, although research universities may have developed relatively late in the South, their recent achievements have given the region a formidable presence in American higher education nationwide.

In June 2011 President Obama introduced the Advanced Manufacturing Partnership, which brought together industry, the federal government, and leading research universities to develop and advance emerging technologies, such as robotics, manufacturing, information technology, biotechnology, and nanotechnology. The president's plan included a $500 million initial budget. Of the six universities selected to participate in the partnership, Georgia Institute of Technology is one, and the manufacturers initially involved will be Allegheny Technologies, Caterpillar, Corning, Dow Chemical, Ford, Honeywell, Intel, Johnson and Johnson, Northrop Grumman, Procter and Gamble, and Stryker.

NANCY SMITH MIDGETTE
Elon College

JOHN R. THELIN
University of Kentucky

AMY E. WELLS
JAMES G. THOMAS JR.
University of Mississippi

George Daniels, *American Science in the Age of Jackson* (1968); Michael Dennis, *Lessons in Progress: State Universities and Progressivism in the New South, 1880–1920* (2001); Dan R. Frost, *Thinking Confederates: Academia and the Idea of Progress in the New South* (2001); Roger Geiger, *Research for Relevant Knowledge: American Research Universities since 1947* (1989); John C. Greene, *American Science in the Age of Jefferson* (1984); Brooke Hindle, *The Pursuit of Science in Revolutionary America, 1735–1789* (1956); Thomas Cary Johnson Jr., *Scientific Interests in the Old South* (1936); Melissa Kean, *Desegregating Private Higher Education in the South: Duke, Emory, Rice, Tulane, and Vanderbilt* (2008); Sally Gregory Kohlstedt, *The Formation of the American Scientific Community: The American Association for the Advancement of Science, 1848–1860* (1976); Nancy Smith Midgette, "The Role of the State Academies of Science in the Emergence of the Scientific Profession in the South, 1883–1983" (Ph.D. dissertation, University of Georgia, 1984); Alexander Oleson and Sanborn C. Brown, eds., *The Pursuit of Knowledge in the Early American Republic: American Scientific and Learned Societies from Colonial Times to the Civil War* (1976); Margaret W. Rossiter, *Women Scientists in America: Struggles and Strategies to 1940* (1982); Amy E. Wells, *History of Higher Education Annual* (2001).

Racialized Medicine

Racial differences in health have long been ascribed to inherited biological differences in susceptibility to disease. In the American South, this tradition can be traced back to the pre–Civil War debate about slavery, when medical "science" was used to justify the institution of slavery by invoking the innate inferiority of blacks to whites. When health conditions for southern blacks deteriorated at the end of Reconstruction, scholars increasingly relied on biological determinism (the idea that social conditions are merely a reflection of biology), rather than attributing the decline to social, economic, and political factors that had created a highly structured racial caste society in the South. By the early 20th century, scientific racism had reached an apex. Researchers used a combination of racially biased empirical evidence and theories to support their claim that disproportionately high morbidity and mortality rates among blacks were a consequence of inherent racial deficiencies. Moreover, the practice of racialized medicine not only rendered blacks as second-class citizens with inferior health status but also exploited them through medical experimentation and involuntary sterilization. The Tuskegee Syphilis Study and eugenics disguised as medical science are among the most infamous examples of the manipulation of black health that resulted from the belief in innate racial disparities. Although the scientific racism of the past has long been discredited, the tradition of ascribing racial disparities in health to fixed biological traits persists. In recent years, notions of racial disparities have again risen to prominence in the search for race-based genetic susceptibility to diseases such as hypertension, heart disease, and infant death.

In the 19th century, when many southern doctors studied the physical and medical differences of blacks, the debate over inherited differences was often guided by prejudice and, prior to the germ theory, based on speculation rather than sound theory. Dr. Josiah Clark Nott of Mobile and Dr. Samuel A. Cartwright of New Orleans gave lectures on "niggerology" and published numerous articles during the 1840s and 1850s. "Cachexia Africana" (dirt eating) and "struma Africana" ("Negro" consumption) were considered exclusively black diseases. Cartwright even invented some disorders, arguing that all blacks suffered from "dysaethesia" (compromised oxygen-carrying capacity in the blood), which together with a deficiency of cranial matter explained their inferior intelligence. He further alleged that fugitive slaves suffered from "drapetomania," a mental disease causing slaves to run away. Still, the relative immunity of blacks to malaria and yellow fever, together with a greater susceptibility to intestinal and respiratory diseases, was undeniable and continually debated by southern physicians. Modern science explained many of the medical phenomena surrounding blacks, but at the time of men like Cartwright, whose audience was concerned with proslavery arguments, there was no challenge to a racial medical system that declared blacks to be of a distinct and inferior race.

At the turn of the century, black health worsened as a direct result of Jim Crow segregation policies that required a costly dual health-care system, which destined blacks to poor health. Mortality and morbidity were higher in the South than in the North, and southern blacks were measurably worse off than even poor whites. The principal factors for disease were poverty and discrimination, and blacks had little or no access to white hospitals and access to very few and often deficient black hospitals. Yet popular theories from newly emerging academic disciplines, such as physical anthropology, genetics, pathology, and evolutionary biology, lent scientific authority to preexisting racial views and promoted a medicalization of racism in the Progressive Era American South. Rather than ameliorating social conditions, the medical community blamed biologic and moral inferiority for the dire state of black health. Public health responses ranged from indifference and inaction to inadequate and often racially motivated efforts to curb disease. White supremacists claimed that blacks were an "unfit" race and maintained that in a matter of time the disease-ridden "Negro" would lose the struggle for survival.

Some diseases were considered almost exclusively racial. For instance, blacks were portrayed as members of a "notoriously syphilis-soaked race." From 1932 to 1972, the U.S. Public Health Service oversaw the Tuskegee Syphilis Study in Alabama, which involved more than 500 black men in the late, tertiary state of syphilis. It became the longest nontherapeutic experiment in medical history. The objective of the study was to investigate the serious complications during the final phase of the disease in blacks by withholding any form of treatment, including therapy for unrelated illnesses. The experiment continued even after a cure in the form of penicillin had become available. The organism responsible for syphilis, the stages of the disease, and the complications resulting from lack of treatment were all known prior to the beginning of the experiment, because a late 19th-century study in Norway had already investigated untreated syphilis in whites. Thus, the Tuskegee experiment was based on the belief that there were pathological differences between the races, and the manifestations of the disease itself were expected to be different in blacks than they were in whites.

Similarly, the sickle cell trait and sickle cell anemia have been linked to race since its discovery in 1910. Physicians regularly used it to assign racial identity, either calling into question their diagnosis because no evidence of racial mixing could be found or doubting the "racial purity" of whites burdened with it. Sickle cell, which was predominately found among blacks in the South, prompted physicians to conclude that the disease could determine if an individual was a "victim of biologic inferiority" and represented a "*status degenerativus*." Thus, sickling was not merely a racial marker but became synonymous for "subnormal" and pathological. In addition, the notion that the trait proved racial distinctiveness between blacks and whites ultimately undermined the perception of "whiteness." If a white person had the trait or disease, it was considered proof of racial mixing. This early 20th-century medical discourse on racial susceptibilities to disease is responsible for the fact that sickle cell continues to be seen as a "black" disease today. Polio, on the other hand, was considered a "white" disease. Cultural and eugenic factors favored a scientific argument that proposed that "primitive" races were less susceptible to a disease afflicting the delicate children of civilized northern Europeans. As a result, many black polio cases were not diagnosed or treated, and inaccurate health statistics further promoted the view of racial differences. Furthermore, white physicians tended to underreport sexually transmitted diseases like syphilis in white patients, as the syphilitic black confirmed racial prejudice about black promiscuity and helped establish the pathological nature of blacks.

Efforts to address the southern black health crisis were further compromised by the prevailing attitude among many white reformers that the best solution would be to reduce the numbers of black births. Public health policy complied with the new discoveries of the laws of heredity by promoting drastic measures such as involuntary sterilization. Extremist racial propaganda, backed by leading scientists, proclaimed the biological and moral inferiority of the "Negro," and these views often invaded the work of white health-care providers. Eugenics was promoted as medical science, and northern philanthropists funded birth control and sterilization programs as part of the public health service throughout the American South, leading to the Negro Project in Nashville, Tenn., and Berkeley County, S.C., in 1939. Margaret Sanger and the American Birth Control Federation envisioned "experimental" clinics that aimed at decimating the black population. General belief in the inferiority of blacks and their presumed inability to use contraceptives led to a halfhearted approach in providing supplies and proper training for birth control, and blacks, desperate to protect their health and limit their family size, often opted for sterilization instead.

Only the advances in microbiology, in particular the notion that disease-causing microorganisms did not differentiate among their victims, led to the gradual development of public health services that addressed and, to some extent, improved the poor health of blacks in the South. Concern for white health could no longer ignore black health, and fear of epidemics and contagious diseases spurred some action out of sheer self-interest. The southern progressive movement was instrumental in the initiation of health reforms to improve southern health across the color line. Yet the degree to which blacks were examined and treated remains debatable. Few black health professionals were ever employed, and, in the age of Jim Crow segregation, forced integration in terms of public health had its clear limitations. The cautious approach, intended to avoid racial strife and open antagonism of southern whites, hampered the extent to which blacks would benefit. Although anxiety over indiscriminate and contagious disease germs had the potential to spur much-needed action, it also served to foster racist ideology by characterizing blacks as notoriously diseased. Public health efforts were successful in reducing the overall death rate, but they also demonstrated the difficulties in terms of racial discrimination. The attitude toward black patients and the unwillingness to accept black leadership in campaigns, such as tuberculosis and hookworm eradication, were largely responsible for the continued disparity in the mortality statistics.

This legacy has shaped the long-term relationship of blacks with the medical profession and public health service. Furthermore, in the wake of the human genome project and race-based genetics, we can observe a renewed scientific legitimacy of racial differences, albeit carefully disguised under the beneficial cloak of "personalized medicine" or "gene therapy." Medical research aims at finding a "race drug" for heart disease or a "preterm birth gene" to explain racial differences in infant death, rather than evaluating socioeconomic factors and minority status as contributors to disease. Even though a purely biological definition for race does not exist among humans, there is a tendency to continue explaining health disparities between whites and blacks as a result of genetic differences. A century ago, a lack of understanding of the transmission of communicable diseases and an eager application of the new (yet poorly understood) gene concept led many to conclude that the disease-causing factors were racial rather than socioeconomic, and most doctors were convinced that blacks were subject to a different pathology.

Today, researchers in population genetics, molecular biology, genomic medicine, and pharmacogenomics claim that naturally occurring polymorphisms have an impact on the frequency of particular diseases in certain populations, and that some racial groups also assimilate drugs differently. Even though none of the genetic components for complex diseases is completely or even largely understood, genetic differences between racial groups often form the starting point in medical research and have already begun to have an impact on clinical utilization and technological application. The use of skin color in diagnostics can result in ineffective treatment and discrimination. Some doctors opt not to treat chronic heart failure with an ACE inhibitor drug in all their black patients, because of studies describing a lower response rate in some black patients. In 2005 the Food and Drug Administration (FDA) approved BiDil, the first drug specifically defined for African Americans, amid great criticism that research failed to support racial variation. In 2007 FDA labeling suggested use of racial information for the dosing of the blood thinner warfarin. Health disparities continue to be discussed in terms of race, and the search for a biological explanation is eerily reminiscent of the biological determinism and its subsequent eugenics movement of a century ago.

ANDREA PATTERSON
California State University

Seymour Garte, *Public Health Reports* (2002); Stephen J. Gould, *The Mismeasure of Man* (1996); James H. Jones, *Bad Blood: The Tuskegee Syphilis Experiment* (1993); Jonathan Kahn, *Scientific American* (August 2007); Nancy Ordover, *American Eugenics* (2003); Andrea Patterson, *Journal of the History of Biology* 42:3 (2009); Naomi Rogers, *American Journal of Public Health* (May 2007); Johanna Schoen, *Choice and Coercion* (2005); Melbourne Tapper, *In the Blood: Sickle Cell Anemia and the Politics of Race* (1999).

Racism, Scientific

The history of "scientific racism" before the 20th century is synonymous with the development of the modern scientific study of race. Scientific racism was not "pseudoscience" but an integral part of the intellectual worldview that nurtured the rise of modern biology and anthropology. In the 20th century, the paradigm of racial hierarchy based on comparative anatomy came under withering attack from the American anthropologist Franz Boas and his students, but, in the history of race science before the emergence of the Boasian school, almost all the participants were racists, and the insights into human diversity provided by the "culture concept" were not available.

Southerners have always had a strong interest in the scientific discourse on race because the fate of their region has been inextricably tied to questions about the role and capacity of African Americans. According to historian Winthrop D. Jordan, Thomas Jefferson's *Notes on the State of Virginia* (1786) was the strongest formal argument for black inferiority published by any native American before the 19th century. Jefferson's work appeared during the period when prejudice against people of color first became a topic of conscious concern among American intellectuals. Jefferson's description of his native habitat was representative of the international effort by natural philosophers to study systematically the bewildering diversity of plants, animals, and peoples revealed by European expansion. Since Aristotle, Western thinkers had found the metaphor of a hierarchical "great chain of being" useful, and many 18th-century natural philosophers expressed their ethnocentric condescension toward Africans by charting pyramids in which European men stood at the apex and the black man below, close to the ape. Jefferson's argument for black inferiority differed from the hardened racism of the 19th century in his regretful and equivocal tone, but *Notes* is one of the documents that marks the general abandonment of the Enlightenment hope that all men could achieve "civilization."

By later standards Jefferson's evidence was "soft." He claimed that blacks were less beautiful than whites, judged their emotional life less complex, and compared them unfavorably with Roman slaves, among whom he found many leaders in the arts and sciences despite harsh conditions that he imagined exceeded those endured by African Americans.

Jefferson argued from history and personal experience, but the major 19th-century American contributions to scientific racism depended on advances in classification and morphology. The taxonomic methods that were serving the botanist well might also be used to explain why Cherokees, Mexicans, and African Americans were fated to serve the "Anglo-Saxon race" or disappear. Assuming that human diversity resulted from differences in heredity, a new generation of natural philosophers found great differences in the skulls of white people and black or red people and turned the abstraction of racial type into a fact of nature.

The reification of race through the taxonomic method is well illustrated in the work of the "American School" of anthropology, whose founder, the Quaker physician Samuel G. Morton (1799–1851), developed the first extensive quantitative data in support of "polygenism"—the theory that human races were separate biological species, the descendants of different Adams. Drawing on a collection of more than 1,000 human craniums supplied by a vast network of correspondents, Morton published a series of works between 1839 and 1849 that were distinguished by brilliant lithographs of skulls and ingenious measurements of their cavities. He argued that a ranking of races could be established objectively through anthropometric measures, particularly brain size, and, unlike Jefferson, he found the Indian as well as the African absolutely inferior to the white in cranial capacity. In 1981 the Harvard scientist Stephen Jay Gould demonstrated that Morton's statistics were a "patchwork of fudging and finagling" but found "no evidence of conscious fraud."

During Morton's lifetime his work was accepted as a model of methodological sophistication and won an international audience. Morton converted Harvard's Louis Agassiz, the dominant figure in American natural science, to the doctrine that races were separate creations but found his most effective disciple in the Alabama physician Josiah C. Nott (1804–73). Nott's *Types of Mankind* (1854; written with George R. Gliddon) provided the authoritative American text on racial differences until Darwin's work necessitated revision of the racist typology. Nott's argument that races were fixed types "permanent through all recorded time" was intended as a rebuttal of abolitionists and racial equalitarians. With the help of his friend and publisher James D. B. De Bow, Nott enjoyed great notoriety and success, despite the hostility of some southern leaders because of his anticlericalism and the incompatibility of polygenism with religious orthodoxy.

Before the Civil War, leadership in scientific racism had passed to Europe, where Paul Broca (1824–80), the French surgeon best known for his discovery of cortical localization in the brain, established himself as Morton's chief scientific heir through painstaking comparative studies of brain weights. By Broca's death in 1880, scientific racism had a well-developed paradigm based on the reification of ideal types and a formidable database drawn from not only skulls and brains but the cranial measurements of 25 million living Europeans. Although the experts were unable to agree on exactly how many races there were or to produce a living example of any pure type, the faith that these types existed shaped such influential popularizations as William Z. Ripley's *The Races of Europe* (1899) and Madison Grant's *The Passing of the Great Race* (1916).

Thus, white southerners had little need to conduct basic research to justify the competitive racial caste system that emerged after the abolition of slavery. The major southern contributors to the scientific literature on race were physicians who described blacks as a diseased and debauched population that would probably be unable to survive without the paternalism of slavery. The medical claims that liberty would lead to black genocide echoed the antebellum myth of a relentless "Anglo-Saxon race" as the agent of the westward march of civilization, bound to exterminate all other breeds that it did not enslave; but the physical degradation of blacks also fit well with the varieties of Social Darwinism that were becoming fashionable.

Although Charles Darwin was a monogenist, the theory of evolution through natural selection proved compatible with the racial typology established by polygenists. Instead of advocating a series of separate acts of creation, late 19th-century scientific racists artlessly worked the established racial types into explanations of human variation that required longer time and gradual change but still assumed that racial types were the ancient determinants of human history. By the end of the century, southern physicians had produced a torrent of abuse in the guise of biomedical studies and had attributed malnutrition, infection, and insanity to a lack of black "fitness" in the struggle for existence. The confidence of white America that the black problem would be solved through extinction of the inferior race was exemplified in the work of Frederick L. Hoffman, a statistician for the Prudential Insurance Company of America, whose *Race Traits and Tendencies of the American Negro* (1896) helped convince most insurance companies that blacks were unacceptable risks.

With white supremacy firmly established in the South by the turn of the century, America's racial anxieties were expressed in campaigns against mass migration from southeastern Europe, and the southern "Negro problem" got relatively little attention. The major development in 20th-century scientific racism in the English-speaking world was the rise of the eugenics movement and its campaigns for sterilization of "defectives," racially discriminatory immigration policies, and tracked school curricula based on intelligence tests. Because most of the immigrants settled in the North and the South's schools were already segregated, the South contributed relatively little to eugenics except for illiterates, dirt eaters, pellagrins, and syphilitics, who served as objects of northern science and philanthropy.

The great public health campaigns that made dramatic contributions to the health of the region were sometimes influenced by scientific racism, most notably in the refusal of many southern physicians to admit that poverty was a better explanation for pellagra than hereditary defect. Many Americans were shocked in 1972 by newspaper headlines describing a U.S. Public Health Service experiment in which more than 400 Macon County, Ala., black men were denied treatment for syphilis as part of an experiment to compare the effects of the disease on blacks and Caucasians. The nightmare that became known as the Tuskegee Syphilis Study had its origins in the effort of northern philanthropists to develop model health programs during the 1920s. When the Great Depression erased the funds for a syphilis-treatment project among blacks in Macon County, government scientists decided to salvage something by charting the natural history of syphilis among sharecroppers who were told that they were being treated for "bad blood." The experiment continued until 1972, when public exposure forced the government to reexamine its policies concerning experimentation with humans. In 1974 the United States agreed to pay approximately $10 million to the victims, but there was no public contrition from the scientists involved. Historian James H. Jones concluded: "Had they been given an opportunity to retrace their steps, there is little doubt they would have conducted the experiment again."

The history of scientific racism exemplifies the powerful influence of social values on the development of biomedical science. White southerners have justified their social institutions in the idioms available to them. When revealed religion provided the primary explanation for the social order, white southern leaders looked to their Bibles; in the 19th century, when science emerged as an important source of authority, white southerners provided an eager audience and offered empirical studies for the new racist science. Long after it had been discredited by the advance of knowledge, racist science influenced medical opinion and helped to legitimate racial injustice. In the case of the Tuskegee Syphilis Study, a measure of justice was achieved not through the initiative of the scientific community but as the result of a lawsuit instituted by Fred Gray, a black attorney and native of Alabama, who first gained prominence in 1955 by defending Rosa Parks for refusing to relinquish her bus seat to a white man. Racism had become a liability for the scientist because those alleged to be inferior had endured long enough to command justice.

JAMES REED
Rutgers University

George M. Fredrickson, *The Black Image in the White Mind: The Debate on Afro-American Character and Destiny, 1817–1914* (1971); Stephen Jay Gould, *The Mismeasure of Man* (1981); John S. Haller Jr., *Outcasts from Evolution: Scientific Attitudes of*

Racial Inferiority, 1859–1900 (1971); James H. Jones, *Bad Blood: The Tuskegee Syphilis Experiment* (1981); Winthrop D. Jordan, *White over Black: American Attitudes toward the Negro, 1550–1812* (1968); William Stanton, *The Leopard's Spots: Scientific Attitudes toward Race in America, 1815–1859* (1960); Nancy Stepan, *The Idea of Race in Science: Great Britain, 1800–1960* (1982); George W. Stocking Jr., *Race, Culture, and Evolution: Essays in the History of Anthropology* (1968); William H. Tucker, *The Funding of Scientific Racism: Wickliffe Draper and the Pioneer Fund* (2002).

Science and Religion (Evolution vs. Creationism)

At the beginning of the 19th century, southern theologians and the region's educated clergy entertained optimistic hopes for an alliance between science and religion. They believed that scientific discovery would confirm theological orthodoxy and even improve the methods of theology itself. By the beginning of the 20th century, that earlier confidence had eroded, and religious conservatives led a series of movements against the teaching of evolution in public schools that still continued in the 21st century. In large part, the change resulted from growing popular awareness of Darwinism, but it also reflected preconceptions formed during the antebellum period.

When John Holt Rice became in 1824 the first professor of theology in the Presbyterian Seminary at Hampden-Sydney, Va., he was officially charged with the task of raising up a generation of scientifically minded clergymen: "That branch of knowledge should form a part of that fund of information, which every minister of the Gospel should possess." The charge embodied a consensus among educated clergymen of the Old South, who were convinced that the scientific investigation of the created order disclosed the existence and nature of the Creator and that the theologian who knew something about natural science could, in the words of Thomas Ralston, "see God . . . mirrored in his works."

The confidence in natural science was an extension of an ancient tradition of natural theology. The antebellum southern theologians—like their northern counterparts—argued that scientific investigation, properly conducted, provided a vast and grand amplification of the traditional argument that design and order in nature demonstrated the reality and trustworthiness of God, and they admired such naturalists as Hugh Miller in Scotland, who had argued that nature was filled with pattern and regularity and therefore with divine intelligence. Southern theological journals carried scores of articles throughout the antebellum period designed to show the harmony of science and religion, and denominational colleges developed courses in chemistry, natural philosophy, geology, and astronomy in order to promote "the cause of science and religion." For the clerical elite, to become an amateur scientist was to extend and enrich the ministerial calling, and few professional groups in the South exhibited greater enthusiasm for the program of natural science than did the educated antebellum southern clergy.

The clergymen did insist, though, that scientists remain within the confines of true scientific method, which they associated with the inductive restraint of Sir Francis Bacon. They opposed scientific materialism, they disliked developmental theories, and they worried about the harmony between science and Genesis. Hence they criticized such scientific skeptics as Thomas Cooper in South Carolina. But most antebellum scientists were themselves Christians, and a few clergymen—such as the Lutheran John Bachman—were respected scientists or amateur naturalists. Most educated clerics had little difficulty reconciling Genesis and geology; they simply argued that the seven days in the biblical creation narrative were geological periods, or they assumed that the creation account in Genesis merely described the final stage of a longer creation. So enthusiastic were they, in fact, that some southern theologians hoped to model theology after the image of the natural sciences.

The 1850s brought signs of strain. The increased interest in developmental hypotheses after the publication in England in 1844 of *The Vestiges of the Natural History of Creation* troubled some clergymen. When the Tombecbee Presbytery in Mississippi recommended the establishing of chairs in theological seminaries to refute infidel naturalists and evince the harmony of science and scripture, some ministers favored the plan because they still expected to maintain cordial relations between the disciplines. The leading proponent, James Lyon of Columbus, Miss., felt confident that God's revelation in nature, deciphered by science, was fully as authoritative and inspired as the Bible itself. When a member of Lyon's congregation, Judge John Perkins, donated $50,000 in 1859, the denomination promptly called James Woodrow to the Perkins Professorship at Columbia Seminary. His inaugural address spoke of the "harmony" of science and scripture. Others, however, supported the plan as a defense against scientific threats.

Feeling pressure from these conservatives, Woodrow soon shifted his language to refer simply to the absence of contradiction between science and religion, and in 1886 he underwent a heresy trial occasioned by his acceptance of evolutionary theory. By that time, the churches were increasingly edgy about the new biology: the Methodists had removed Alexander Winchell from a post at Vanderbilt in 1878 partly because of his defense of Darwin. But when statewide struggles over the teaching of evolutionary theory in the public schools erupted in the 1920s, the churches lined up on both sides of the issue.

The opponents of evolution were noisy, but only in the Southern Baptist Convention did they succeed, in 1926, in securing an official condemnation of developmental theories of human origins. Antievolutionary groups also had mixed success in the state legislatures and statewide referenda, passing laws against the teaching of evolution only in Florida, Tennessee, Mississippi, and Arkansas. William Jennings Bryan's appearance at the trial of John Scopes in 1925 strengthened the antievolutionary sentiment in the rural South, but by the end of the 1920s the issue seemed to fade away. Prominent religious leaders, including the Baptist E. Y. Mullins, the Methodist bishops E. D. Mouzon and John M. Moore, and the Presbyterian Hays Watson Smith, had openly opposed the antievolutionary movement, and the opponents of Darwin had failed with the voters and legislatures in all but four states.

Some of the religious opposition to Darwinism reflected a broader opposition to science itself, but many of the fundamentalist opponents of Darwin claimed to be friends of science. They contended only that Darwinism had transcended the bounds of scientific method propounded by Francis Bacon: the evolutionists, they claimed, were insufficiently inductive. Hence with the resurgence of scientific creationism in the 1960s—stimulated in part by the writings of Henry Morris of Virginia Polytechnic Institute—southern religious conservatives again tried to compel the schools to teach a biblical science, and only a federal judicial decision in 1982 thwarted them in Arkansas.

The term "theory"—and, even more, a distinction between "theory and fact"—had long assumed a prominent place in the debates, and the 1980s brought a shift in strategy among the opponents of evolution. In 1980 Texas adopted policies stating that evolution should be taught "only as a theory," not as fact. In 1992 the Alabama State Board of Education inserted into biology textbooks the disclaimer that "any statement about life's origins should be considered as theory, not fact." In 1995 Tennessee legislators debated a bill that would have allowed school systems to fire teachers who taught evolution "as fact rather than theory." During the 1990s serious controversies regarding evolution and creation disrupted secondary education in 14 states. In 2004 the state school superintendent in Georgia wanted to remove from the proposed biology curriculum for the state's schools any reference to the word "evolution" and replace it with the phrase "biological change over time" as part of her effort to expose the state's students to "all legitimate theories." The Cobb County school system, one of Georgia's largest, pasted a disclaimer in biology textbooks repeating the "fact-theory" distinction, but a judicial order in 2006 forced its removal.

By that time, many opponents of evolution had shifted their vocabulary to speak of "intelligent design" rather than "creationist science," but the aim was still to counter the absence of teleology in Darwinian and neo-Darwinian science and to present an alternative more in accord with a theological view of the created order. In 2000 disputes over the teaching of intelligent design generated conflict at Baylor University in Texas, which disavowed its promotion at the affiliated Michael Polanyi Center for Complexity, Information, and Design.

Antievolutionary ideas reflecting religious beliefs continued, however, to attract widespread support in popular culture. In 2007 opponents of evolution, funded by donations to an antievolutionary ministry, built an elaborate Creation Museum near Petersburg, Ky., which claimed a million visitors by 2010. Its exhibitions tried to display graphically the harmony between a biblical account and a scientific understanding of natural and human history. Similar smaller museums drew viewers in Texas, Florida, Tennessee, and Georgia. In 2010 the Gallup Poll found that 40 percent of Americans believed that God created human beings in their present form within the past 10,000 years, but the percentage was higher in the South, where only 27 percent accepted Darwinian arguments. Not all religious conservatives accepted this version of a "young earth" theory, but most of them still rejected Darwinian views. In other words, religious positions had hardened since the 19th century, when several conservative southern theologians had no difficulty interpreting Genesis as referring to geological periods or a long period of creation prior to the events of the biblical six days. It is hard to know how widely such ideas circulated among people with little or no education.

The South is now home, however, to several liberal and moderate schools of theology in which theologians easily accept some variant of a neo-Darwinian evolutionary biology. In Catholic, Presbyterian, United Methodist, moderate Baptist, and Episcopal seminaries, a form of biblical criticism that insists on the symbolic and religious character of the creation accounts in Genesis is taken for granted. In such institutions, as well as in the denominational colleges of these churches, most faculty members accept an evolutionary development of the natural and human world, though some still have reservations about the absence of teleology in the mainstream of neo-Darwinian thought. Some argue that a nonteleological metaphysic is not essential to a Darwinian account, others simply live with the tension between a theology that affirms purpose and a science that views natural selection as accidental, and still others argue that theology presents a metaphorical account of the meaning of human life rather than a scientific description of its origins. The lay members of these denominations, however, are more divided, with the conservative view enjoying a strong representation, especially outside the large cities. In the largest Baptist seminaries, as well, especially since a fundamentalist takeover in the 1980s, strong anti-Darwinian views still prevail, and most Southern Baptist lay people share those views.

Both the 19th-century proponents of natural theology and the opponents of Darwin usually insisted that the Bible was both a religious and a scientific text and that religious and scientific assertions were therefore equivalent in logical status. The 21st-century southern opponents of evolutionary science have held on to the quasi-rationalistic presuppositions of the older 19th-century theological orthodoxy. But southern religion is now far more diverse than it was at the beginning of the 20th century, and so are views about religion and science.

E. BROOKS HOLIFIELD
Emory University

Theodore Dwight Bozeman, *Protestants in an Age of Science: The Baconian Ideal and Antebellum Religious Thought* (1977); Norman Furniss, *The Fundamentalist Controversy, 1918–31* (1954); Willard B. Gatewood Jr., *Preachers, Pedagogues, and Politicians: The Evolution Controversy in North Carolina, 1920–1927* (1966); Langdon Gilkey, *Creationism on Trial: Evolution and God at Little Rock* (1985); E. Brooks Holifield, *The Gentlemen Theologians: American Theology in Southern Culture* (1978), *The Odd Couple: Theology and Science in the American Tradition* (2005); Edward L. Larson, *Summer for the Gods: The Scopes Trial and America's Continuing Debate over Science and Religion* (1997); David C. Lindberg and Ronald L. Numbers, eds., *When Christianity and Science Meet* (2003); George Marsden, *Fundamentalism and American Culture: The Shaping of Twentieth-Century Evangelicalism, 1870–1925* (1980); James R. Moore, *The Post-Darwinian Controversies: A Study of the Protestant Struggle to Come to Terms with Darwin in Great Britain and America, 1870–1900* (1979); Ronald L. Numbers, *The Creationists: The Evolution of Scientific Creationism* (1992), *Darwinism Comes to America* (1998), *Science* (5 November 1982); Jon Roberts, *Darwinism and the Divine in America: Protestant Intellectuals and Organic Evolution, 1859–1900* (1987).

Slavery and Medicine

Early histories that focused on slavery and medicine in the American South had typically examined the enslaved as either hapless victims or practitioners of hoodoo and conjuring. In the same spirit, white southern antebellum-era physicians have also been characterized by myopic renderings that have painted them as either saints or sinners in their treatment of black people. More recent scholarship on these themes, however, has broadened intellectual discussions to include the perspectives of enslaved women and the poor and to examine midwifery as a skilled trade. These new offerings have even explored the political dimensions of southern medicine and slavery as a major point of contention in the bitter regional conflict that became the Civil War. Most critically, the outpouring of writings on the South, slavery, and medicine have nuanced current understandings of the role that slavery played in the development of medicine in the South, the nation, and the 19th-century global world.

Globally, the medical advancements pioneered by antebellum-era southern physicians have altered, permanently, two branches of medicine: gynecology and epidemiology. Innovations in gynecological surgery led to the repair of obstetrical fistulas, and the application of insect vector theory to yellow fever reduced drastically its death rate. Although free black and enslaved populations were overused and sometimes exploited in the name of medical experimentation, some southern white physicians blazed trails in several medical fields during the long 19th century. Additionally, black midwives, who were the slave counterparts to these medical men, had overwhelmingly successful records in the preservation of their patients' lives despite the immeasurable odds they faced, especially in a pre-germ-theory world.

It is vital to understand the history of how slavery and American southern medicine were first linked and dependent upon each other. Although Massachusetts, a northern colony, first legalized slavery in 1641, southern colonies came to depend heavily on the labor of black slaves by the end of the colonial era. Even as a nationalizing project, the agricultural labor of the enslaved helped to galvanize the American South's economy via cash crops like tobacco, sugarcane, and cotton. By the start of the 19th century, the American South was firmly a slave society. Despite the lucrative nature of slavery, it was a costly undertaking to care for millions of black bondsmen and bondswomen. White slave owners were confronted with numerous challenges around the issue of slave medical care. Most of the slave management advice dispensed by slaveholders and overseers in farm journals focused on the health care of the enslaved. The maintenance of a healthy slave work force ensured the successful continuation of the institution of southern slavery.

One Mississippi physician and slave owner wrote in a farm journal in 1846, "Were you to give me the selection of any one subject . . . for the express design of benefitting my family and country, I know not but what I should name—the treatment of our slaves. I mean not only as regards their labor, but entire treatment, whether in health or sickness." Thus the growth of modern American medicine and slavery had extensive roots in the South. In slave markets dotted throughout the southern United States, physicians were a ubiquitous sight. They were there to administer medical examinations to the enslaved, and they helped to determine the soundness (health) and, subsequently, the economic value of bondsmen and bondswomen. Despite the presence of southern white male physicians, the enslaved were also involved in medicinal practices where they doctored themselves by "working cures" or utilizing root and herbal-based homeopathic remedies.

Enslaved women provided the health-care needs of their community members. Slave healers created pharmacological concoctions that were not invasive and administered them with a spiritual mix that had the remnants of Western and Central African spiritual practices. Unlike the formal and sometimes scientific medicine practiced by southern white physicians, enslaved men and women viewed folk medicine, also called hoodoo, as a healing art that was derived from nature and imbued by a higher supernatural power.

The practice of folk medicine was not the sole domain of black women healers but was also relied on by white women, including plantation mistresses. The labor of both groups of women was representative of the racialized and gendered politics of the American South. Medicine, even folk medicine, operated in two spheres—one domestic and the other institutional. Yet southern white men chose, increasingly, to enter into medical schools and practice in new fields of medicine like obstetrics and gynecology. A number of these southern men had to travel either to the North or to Europe to matriculate at medical colleges because the South housed just five medical schools by the 1840s. Many of the antebellum-era's pioneering doctors in professional women's health were southerners. The access they had to black female slave bodies in particular aided them in their burgeoning careers as gynecological surgeons.

The best-known gynecologist of the era, Dr. J. Marion Sims, a native of South Carolina, created the surgical cure for obstetrical fistulas, a common 19th-century disorder that affected pregnant women after delivery. Sims did so by conducting experimental surgery on bondswomen he either owned or leased over a four-year period in Alabama. After years of failed surgeries, he taught his enslaved patients to serve as surgical nurses for each other during the final phase of his medical experimentation. After achieving national and international fame, Sims relocated to New York to train other physicians in this kind of surgery. Because of his medical achievements, Sims is regarded as "The Father of Gynecology."

Sims's professional successes advanced modern American gynecology, but what is also central to examine is how the institution of southern slavery aided his medical research. Equally important is understanding how crucial the reproductive labor of enslaved women was to Sims as it helped the surgeon to improve gynecological surgery. The Sims historical example is also representative of the saliency of southern slave labor to a burgeoning national medical industry. The availability of black bodies and, to a lesser degree, poor white bodies allowed doctors and surgeons like Sims to perfect their crafts. In their medical treatment of the enslaved, white slaveholders and physicians could extol the then-prevailing paternalistic notion of the "benevolent" white father. The white, southern father archetype was propagated throughout the South to demonstrate how much slave owners cared for "their" black children. This politicized display of white southern magnanimity, which focused on the medical needs of the enslaved, was discussed in proslavery propaganda and became fodder for political leaders who resided on opposite sides of the slavery and abolition spectrum.

By the end of the Civil War, the medical field was fragmented by racial animosity, acrimonious political battles, and the introduction of segregationist Jim Crow laws. Buoyed by their newfound freedom, formerly enslaved men and women looked to the Medical Department of the Bureau of Refugees, Freedmen, and Abandoned Lands (Freedmen's Bureau) to provide an easy transition from slave-sanctioned health care to independent medical care. Unfortunately, freedmen and freedwomen in cities like Memphis, Tenn., and Athens, Ga., who suffered from wartime wounds, small pox, malaria, and cholera died in massive numbers. Their demise was caused by the negligence of a medical system stymied by political infighting between bitter local officials who resented the intrusion of national officials from the North. Further, the Freedmen's Bureau's hospitals and medical centers did not often have the infrastructure to accommodate the massive numbers of indigent black poor who needed medical treatment.

The legacies left by slavery and southern medicine are seen in the urban hospitals in cities like New Orleans and Charleston, S.C. Many, like the New Orleans Touro Infirmary, established in the 1850s, still serve a primarily poor and black population much as they did during the era of slavery. The gynecological work of 19th-century physicians such as Ephraim McDowell, François Marie Prevost, and J. Marion Sims, who relied heavily on the bodies of enslaved black women, pioneered surgeries like the ovariotomy, Caesarean section surgeries, and the repair of obstetrical fistulae. Their surgical techniques are still in use today. Fortunately, scholars of medicine, slavery, gender, and race have complicated outdated notions of southern exceptionalism and moved the

South from the margins to the center in national conversations about these categories.

DEIRDRE COOPER OWENS
University of Mississippi

James O. Breeden, ed., *Advice among Masters: The Ideal in Slave Management in the Old South* (1980); Sharla M. Fett, *Working Cures: Healing, Health, and Power on Southern Slave Plantations* (2002); Peter Kolchin, *American Slavery, 1619–1877* (1993, 2003); Stephanie Y. Mitchem, *African American Folk Healing* (2007); Martin S. Pernick, *A Calculus of Suffering: Pain, Professionalism, and Anesthesia in Nineteenth-Century America* (1985); Todd L. Savitt, *Race and Medicine in Nineteenth- and Early-Twentieth-Century America* (2007); Marie Jenkins Schwartz, *Birthing a Slave: Motherhood and Medicine in the Antebellum South* (2006).

Slaves in Medical Education and Medical Experiments

In his autobiography, *Slave Life in Georgia*, fugitive slave John Brown described his experience as a human guinea pig in a series of distressing medical experiments suffered at the hands of physician-planter Dr. Thomas Hamilton. Having successfully cured slave owner Thomas Stevens, Hamilton was granted the favor of borrowing one of his slaves, Fed (as Brown was then named), "for the purpose of finding out the best remedies for sun-stroke." Brown's important narrative provides a rare insight into how the chattel principle of antebellum southern slavery (the idea and practice of owning other human beings) could render slaves "medically incompetent," unable to determine the course of their own health, or prevent risky and undesired interventions. Brown records that neither he nor his owner knew what the nature of Hamilton's experiments were and, furthermore, that Stevens never cared to ask the physician any questions on the subject. Brown remembered that, "Even if I had been made aware of the nature of the trials I was about to undergo, I could not have helped myself. There was nothing for it but passive resignation, and I gave myself up in ignorance and in much fear." Brown's story is a tragic illustration of how enslaved people struggled to assert their autonomy and will in the face of slaveholder power, as their rights and voices were too often silenced by the indifference of slave owners and the professional interests of physicians.

In the mid-19th century knowledge of anatomy and the practice of dissection were seen as an essential component of a physician's apprenticeship, as well as being central to programs of professional medical education. Many northern cities and medical institutions, facing public hostility and legal restrictions to the study of practical anatomy, struggled to furnish medical students with a sufficient supply of cadavers. By contrast, medical schools in major southern cities and slave-trading centers, such as Augusta, Richmond, Savannah, Atlanta, New Orleans, Memphis, and Charleston, boasted that they were able to guarantee adequate and reliable sources of corpses, attracting students eagerly in search of the core skills that would distinguish and legitimate their practice. The faculty of Charleston's Medical College of South Carolina declared that they had access to an abundance of "material" for those seeking knowledge of gross anatomy and wishing to practice surgery. "No place in the United States offers as great opportunities for the acquisition of Anatomical knowledge, subjects being obtained from among the colored population in sufficient number for every purpose, and proper dissection carried on without offending any individual in the community," boasted the College. "Those impediments which exist in so many other places, to the prosecution of this study, are not here thrown in the path of the Student, public feeling being rather favourable than hostile to the advancement of the Science of Anatomy."

Charleston's reserves of dead bodies were so easy to obtain that they supported a number of private schools in the city. In addition to the instruction offered in the city's two main medical colleges, entrepreneurial physicians offered "preparatory" classes and summer institutes, advertising courses in which dissection occupied a central place in the curriculum. Such was the surfeit of anatomical material and the demand for instruction in Charleston that William Middleton Michel was able to develop his own private anatomical school and medical museum at the corner of Franklin and Queen streets. Added to these anatomical resources were the bodies of patients who died in the medical college's Negro Infirmary and at small private slave infirmaries operating within the city. Students deciding to study medicine and practice anatomy in Charleston and similar medical centers across the South did so with the assurance that their exploitation of slave cadavers was unlikely to trouble the public conscience.

The acquisition of slave patients and slave bodies for medical research and education was first explored in-depth by historian Todd Savitt in *Medicine and Slavery* (1978) and is also examined by Blakely and Harrington's *Bones in the Basement* (1997), an edited collection of essays published following the archaeological excavation of the original Medical College of Georgia building in Augusta. Savitt noted that there were multiple sites and circumstances where enslaved people could be exploited for medical instruction and experimentation, including medical colleges, infirmaries, physicians' offices, physicians' presentation of cases to medical society meetings, and physicians' submission of cases to medical journals and medical museums. All of these agents and institutions utilized anatomical material and experimental subjects for purposes of education or research, and Savitt's evidence for antebellum Virginia indicates that enslaved patients and bodies played an essential role in fulfilling the demands of the developing profession.

As in Charleston, opportunities for practical anatomical instruction in Augusta, Ga., received special mention in the Medical College's annual announcements—"the whole Faculty . . . have made the most reliable arrangements for an ample supply of material." These arrangements were in large part the result of the college having acquired the services of an enslaved man, Grandison Harris, to labor as a resurrectionist and dissection-room porter. A powerful Gullah slave bought "from the auction block" in Charleston by the dean in 1852, Harris was prized by the college's anatomists for his ability as a "sack-'em-up man," or body snatcher. Following Harris's arrival in Augusta, the number of corpses purchased for gross anatomical studies took a dramatic drop, as he was supplying the college by raiding local slave cemeteries. Blakely and Harrington's archaeological analysis revealed that the college basement was used to dispose of the dissected remains of cadavers throughout the 19th century, and the researchers determined that a majority of the bones discovered there, beneath the floor of what had been the anatomical theater, were of African American origin—evidence of anatomical or "postmortem racism," a practice widespread throughout antebellum southern medical education.

Anatomical material for medical instruction and research also found its way to southern medical students and researchers through networks connecting college, hospital, journal, and rural hinterland. Anatomical and pathological specimens often accompanied the submission of case-history narratives to medical journals and many specimens were deposited in college medical museum collections. Case histories published in southern medical journals indicate that a large number of body parts were appropriated from enslaved African Americans for antebellum southern medical museums and private collections. While practicing in the cotton plantation district of Quincy, Fla., the Louisville-educated physician Charles Hentz recorded in his diary having taken home specimens acquired from slave patients. A three- or four-month-old fetus was retained for its pristine appearance, along with the dried and varnished leg bones of a male slave who died following surgery. The acquisition and retention of these specimens were greatly assisted by the power even relatively isolated white southern physicians commanded over slave bodies.

As African American slaves were legally considered to be commodities and framed as an inferior race by white slaveholding ideology, the southern medical profession faced few obstacles utilizing slave patients and their bodies for teaching and research. Southern physicians, such as Thomas Hamilton and J. Marion Sims, were well placed to exploit their power and authority over slaves by conducting operations and experiments that would have proved tremendously difficult to perform on white patients. An awareness of the developing market in slave medicine encouraged enterprising antebellum southern physicians to establish their own private black infirmaries. In these environments, with slave patients subjected to close and constant medical supervision, southern doctors were at liberty to develop their skills in surgical science.

In Montgomery, Ala., J. Marion Sims developed a reputation as "a bold, fearless, and dashing operator" able to "undertake almost any case" and became a magnet for chronic infirmities. Slave owners sought out his assistance in last-ditch attempts to preserve their investments, and Sims willingly accepted their requests, converting slave debility into his own regional, national, and, eventually, international medical celebrity. In a case history published in the *American Journal of Medical Sciences* in 1846, Sims recounted an operation he performed on Sam, a 26-year-old enslaved male. Sims diagnosed Sam's ailment as cancer of the lower-jaw and decided that an operation was the only means of saving his patient's life. Sam had vigorously resisted all earlier attempts at treatment, and recognizing that surgery "would never be done with his consent," Sims "determined not to be foiled in the attempt." By means of surgical bandages, Sam was bound to an operating chair so tightly that his legs, body, and hands were immovable. His head was secured using a leather strap, which could be controlled so as to hold it in any position that the surgeon wanted. Sims's will for absolute control of Sam's body highlights that the operation he attempted was hazardous and required working far beyond the range of common medical practice.

In reviewing Sam's case at the close of the article, Sims discloses that the "operation was performed in the presence of a large number of medical gentlemen." At least 25 medical observers (10 medical students and 15 doctors) watched Sims perform this exceptional surgery, while 5 of the 15 doctors present took a more active role in restraining the patient and offered their assistance to the principal surgeon. Given the size of the specialist audience this surgical spectacle attracted, it is clear Sims's slave infirmary also doubled as a private medical research facility. Here, Sims had the opportunity to showcase his talent to fellow and would-be professionals without fear of censure from the Montgomery public. His slave patients functioned as ideal subjects, and consent was not an issue Sims needed to consider when operating upon them. As the physician's paying client, the slave owner and his interests were paramount, whereas the enslaved patient was effectively rendered voiceless and powerless. The second of six points Sims listed in summary made a virtue of this process of domination, stating that Sam's case proved "the practicability of the operation, whether the patient is willing or not."

As cadavers for dissection, anatomical and pathological objects encountered in the learning spaces of medical museums, and as clinical subjects in black infirmaries, African American bodies served as one of the principal means through which southern doctors learned their trade, generated and communicated medical knowledge, and increased their economic wealth along with their social and professional status. African American patients and their bodies offered southern physicians not only opportunities to construct successful everyday practices in a slaveholding economy but also the resources for making professional reputations as active medical researchers. As late 19th-century dissection room photographs and the tragic story of the Tuskegee Syphilis Experiment vividly demonstrate, the exploitation of African Americans for anatomical and experimental purposes outlasted the institution of chattel slavery and survived into the medical practices of the post-Emancipation period and beyond into the 20th century.

STEPHEN C. KENNY
University of Liverpool

Robert L. Blakely and Judith M. Harrington, eds., *Bones in the Basement: Postmortem Racism in Nineteenth-Century Medical Training* (1997); John Brown, *Slave Life in Georgia: A Narrative of the Life, Sufferings, and Escape of John Brown, a Fugitive Slave, Now in England* (1855); Sharla Fett, *Working Cures: Healing, Health, and Power on Southern Slave Plantations* (2002); Todd L. Savitt, *Medicine and Slavery: The Diseases and Health Care of Blacks in Antebellum Virginia* (1978); Marie Jenkins Schwartz, *Birthing A Slave: Motherhood and Medicine in the Antebellum South* (2006).

Surgeons General

Since the late 1800s, the U.S. surgeon general, or "America's Doctor," head of the Public Health Service (PHS), has been advising the American public about its health. The Marine Hospital Service (MHS), precursor to the Public Health Service, was established in 1798, but the first supervising surgeon of the MHS was not appointed until 1871, after an 1870 reorganization of the service as a national hospital system. The supervising surgeon post was filled by a medical officer and appointed by the president and approved by the Senate. The name for the position became "surgeon general" in 1902. Until 1968 the surgeon general was administratively responsible for all activities of the PHS. In that year, the surgeon general's main duty of medical advisement to the public (along with the operational command of the Commissioned Corps of the U.S. Public Health Service) was clarified.

Today, in 2011, the United States has had 18 surgeons general. Of those 18, six were born in the South, and two of the three female surgeons general were southerners. Most of these southern surgeons general received their medical training from schools in the South, and several have listed their particularly southern experiences as formative in their careers, including their tenures as surgeons general.

The first southern surgeon general is actually not included in the count of U.S. surgeons general but made significant contributions to southern military medicine and, subsequently, to national military medicine. Samuel Preston Moore, the longest-running surgeon general of the Confederate States during the Civil War, is best known for his development of "hospital huts," or barracks hospitals, for the Confederate army. The idea soon caught on in northern camps and prompted significant strides in military hospital sanitation across the country.

Moore was born in 1813 in Charleston, S.C., and received his medical education from the Medical School of South Carolina. In March 1835 he became assistant surgeon (with the rank of captain) in the United States Army. During his career as a U.S. military surgeon, he was assigned to posts all over the country, from Florida to Oregon. He practiced as medical purveyor in New Orleans, La., from April 1860 until February 1861, when he resigned from the U.S. Army, just a few months after his home state seceded from the Union (20 December 1860). Two other men served as surgeons general of the Confederate army in the few months before Moore's tenure: David C. DeLeon (of Mobile, Ala.), and Lafayette Guild (of Tuscaloosa, Ala.), who later became medical director of Lee's Army of Northern Virginia. Moore replaced Guild in November 1861 and remained the Confederacy's surgeon general until the end of the war.

During his tenure, Moore organized the Association of Army and Navy Surgeons of the Confederate States of America and, for a year (1864–65), published the *Confederate States Medical and Surgical Journal* from Richmond. Moore was responsible for shaping the chaotic Confederate army medical corps into a much more efficient organization. He raised recruitment standards by requiring board examinations and training for physicians who wanted to enter the corps, and he standardized treatment protocols for Confederate hospitals. He also led a successful effort to find southern plant substitutes for pharmaceuticals that were scarce owing to Federal blockades of Confederate transportation and shipping lines. Moore's most remembered contribution, however, was the development of the aforementioned design for barracks hospitals. In response to the extreme shortage of sanitary medical facilities, Moore developed a plan for easily constructible "hospital huts," or wards, that held upward of 32 patients each and allowed for standardization of medical treatments. These barracks hospitals were used for the first time in the Confederate army and led to the establishment of five general hospitals (of three divisions each, each containing 15 to 20 "huts," which could contain 40 to 60 patients) in Richmond—the largest military hospital in the world at the time.

The first southern U.S. surgeon general would not take office until many years later. Rupert Blue, a North Carolinian, was appointed as fourth United States surgeon general in 1912. He served in that post until 1919 and presided over an extensive restructuring and expansion of the Public Health Service, while adapting to challenges presented by World War I. Before his tenure as surgeon general, Blue was assistant surgeon with the Marine Hospital Service. He is probably best known for his achievements during this time, when he was dispatched to San Francisco several times to develop and implement rat and squirrel eradication measures after the 1906 San Francisco earthquake and fire. He proved that rats were carriers of the bubonic plague that ran rampant in the city, and he effectively stopped the plague in the city in 1908. In addition to sanitation projects in South America, Blue directed more disease-containment efforts with mosquito eradication measures in New Orleans, La., in 1905, in Jamestown in 1907, and Honolulu and the Panama Canal in 1911.

After he became surgeon general, though, Blue continued the same sorts of disease-containment measures (including a 95-day eradication of bubonic plague in New Orleans in 1915). Along with the physicians and researchers under him, Blue began working to control diseases that they believed were linked to poverty and poor sanitation in rural and urban areas. The worst representations of the causative effects of poverty and ill health appeared most starkly in the South, so many of Blue's efforts were applied there. Under his direction, Dr. Charles Stiles studied hookworm disease and its connection to contaminated soil in the U.S. Southeast, and Dr. John McMullan worked to develop a treatment for trachoma in communities in the Ozarks and southern Appalachian mountain regions (1913).

Blue also assigned Dr. Joseph Goldberger to the South to study and develop treatments for the pellagra problem there. Pellagra was particularly severe in Mississippi, Arkansas, Alabama, and Georgia, so Blue stated in a 1912 interview that "investigations of pellagra are to be pushed in the Southern states." Goldberger conducted experiments to determine the cause and cure for pellagra in seven cotton-mill communities near Spartanburg, S.C., and in a potentially controversial secret study on prisoners at Rankin Farm in Greenfield, Miss. He and his colleagues were able to induce pellagra and then cure it in healthy white male volunteers simply by changing the subjects' diets. In a report on his findings, Goldberger and his partner, Dr. Edgar Sydenstricker, wrote that pellagra and its cause in dietary deficiency were "deeply rooted in the political economy of cotton monoculture in the South." After his tenure as surgeon general, Blue spent several years (until his retirement in 1932) as a U.S. delegate to international health and hygiene boards.

Blue's immediate successor, the fifth U.S. surgeon general, was also a southerner. Hugh S. Cumming, born and educated in Virginia, served as surgeon general from 1920 to 1936. After his medical education at the University of Virginia Department of Medicine and Virginia College of Medicine, Cumming was commissioned as assistant surgeon in the Marine Hospital Service (which, in 1912, became the PHS). During his early career there, he was assigned to several different posts at quarantine stations in the South and on the West Coast. From 1913 to 1919 he had charge of the Washington, D.C., Hygienic Laboratory's investigation of polluted tidal waters in Maryland and Virginia; he was particularly concerned about health risks to humans who ate contaminated shellfish.

One of the earliest events of Cumming's tenure as surgeon general was his direction of the creation of a national leprosy hospital in Carville, La., in 1921. What had been the "Louisiana Leper Home" became a major center for research on and treatment of leprosy. Interestingly, the new Carville Leprosy Hospital, which remained in operation until 1999, did not follow the precedent for many southern facilities in the pre–civil rights era; Carville, from its beginning in 1921 until its closure in 1999, was never segregated.

The blight on Cumming's career as surgeon general was the beginning of the infamous Tuskegee Syphilis Study. His administration began it in 1930, and the study, which examined the results of untreated syphilis in African American males, lasted until 1973. In a 2006 article for the *Bulletin of the History of Medicine*, Paul Lombardo and Gregory Dorr speculated that Cumming's training at Virginia medical schools may have had something to do with both his official involvement in the racialized medicine of the American Eugenics Movement and his condoning of the Tuskegee experiment.

After his retirement from his surgeon general post, Cumming became director of the organization that would later become the Pan-American Sanitary Bureau, and he was the longest-running director of that program, which dealt with immigration and quarantinable diseases.

Another southerner would not occupy the surgeon general post until 1961, when President John F. Kennedy appointed Luther Leonidas Terry, a native of Red Level, Ala., to the position. Terry would be the first of three surgeons general from Alabama. The American public probably became most aware of the surgeon general's office during Terry's tenure, for his contribution to the country's public health was far-reaching. He is best remembered for the advocacy that began public awareness about the health risks of smoking tobacco.

Though Terry completed his medical residency outside the South in Cleveland Hospitals, and completed an internship in pathology at Washington University in St. Louis, the bulk of his medical training, like Blue's and Cumming's, occurred in the South—at Birmingham-Southern College and Tulane University. After a pathology internship in St. Louis and a one-year instructor position there, Terry returned to the South and the University of Texas at Galveston as assistant professor of preventive medicine and public health, and his involvement in the public health scene began. Within two years (by 1942), he had joined the Public Health Service Hospital in Baltimore. Soon, because of his cardiovascular research, he became the chief of General Medicine and Experimental Therapeutics at the National Heart Institute. While he was assistant professor from 1944 to 1961 at Johns Hopkins University School of Medicine, the Heart Institute quickly became his niche.

Perhaps his interest in cardiovascular health, along with a 1962 report from the British Royal College of Physicians concerning health and smoking, led Terry to start the surgeon general's Advisory Committee on Smoking and Health to conduct a similar study in the United States. The committee's report, released on 11 January 1964, concluded that lung cancer and chronic bronchitis, and perhaps emphysema and cardiovascular disease, were directly and causally related to cigarette smoking. Terry had himself quit smoking in 1963, and he began urging Americans, as a result of this study, to do the same. In fact, the committee's report so significantly linked smoking to those health problems that the U.S. government and the PHS desired legislative action to regulate the cigarette industry. The American public, alarmed, began antismoking campaigns, and in 1965 the Cigarette Labeling and Advertising Act mandated the now-familiar surgeon general's health warnings that appear on cigarette packages.

Terry left the surgeon general's office in 1965 but continued his work with antismoking campaigns, chairing the National Interagency Council on Smoking and Health (1967–69) and pushing for subsequent legislation, like the Public Health Cigarette Smoking Act (1969) and a ban on radio and television cigarette ads (1971). Terry's work, which laid foundations for subsequent antismoking campaigns, began a cut of steady smokers in America from 43 percent to 22.8 percent (from 1964 to 2001). Terry died in 1985 and was buried in Arlington National Cemetery in Arlington, Va.

The next southerner to hold the post was a landmark surgeon general—the first African American to hold the position and the second woman: Joycelyn Elders, a native of Schaal, Ark. Born in 1933 to a poor farming family, Elders named her rural background as formative in her decision to pursue a medical career. At a sorority meeting at her all-black liberal arts Philander Smith College in Little Rock, she was inspired by a speech by Edith Irby Jones, the first African American to attend the University of Arkansas Medical School. Elders decided to become a doctor after this meeting, and she received most of her graduate medical training at the University of Arkansas. She was the first person to become board certified there in the field of pediatric endocrinology, and, by 1976, Elders held the position of full professor in pediatrics at the University of Arkansas. As she studied endocrinology (growth and sexual development in children and adolescents) over the following 20 years, Elders pointed out a correlation between juvenile diabetes and pregnancy risks in teen mothers. In her medical practice, she began more intensively studying sexual behavior and advocating birth control measures for young women.

Her connection to public health began when Bill Clinton, then governor of Arkansas, appointed her as head of the Arkansas Department of Health in 1987. During her tenure there, Elders campaigned extensively for "expanded sex education," which made sex education and contraceptives more readily available to teenagers, as well as counseling and HIV testing. Though many conservative groups opposed her advocacy, her work resulted in a 1989 mandatory, statewide K–12 sex education, substance abuse prevention, and self-esteem promotion program. Also, her tenure with the Arkansas Department of Health from 1987 to 1992 saw the doubling of childhood immunizations and expansion of prenatal care in the state.

In 1993 President Bill Clinton appointed Elders as surgeon general. Her tenure was short lived, however. She continued her controversial advocacy of mandatory sex education and was also criticized for her outspokenness on marijuana legalization. Elders resigned after 15 months as a result of the controversial statements she made concerning sex education. She returned to her former position at the University of Arkansas after leaving the surgeon general's post and became a professor emeritus at the University of Arkansas School of Medicine. In 1996 Elders published an autobiography that embraced her sharecropping roots in Arkansas, especially the lack of rural primary care there, as beginnings for her career.

The same lack of rural primary care seemed to be a motivator for the next southern surgeon general, David Satcher. This 16th surgeon general was born in Anniston, Ala., in 1941. His childhood in rural Alabama seemed to have been a formative experience for his medical career, shaping his activism of later years and his tenure as surgeon general. At age two, Satcher contracted whooping cough, but because he was African American (and it was the early 1940s) he could not be treated at the nearest (white) hospital. A local black doctor saved his life, and by the age of six, Satcher had decided that he wanted to become a doctor.

Satcher's medical training departed somewhat from that of previous southern surgeons general; after his graduation from Morehouse College in Georgia in 1963 most of his training took place outside the South. Satcher received his medical degrees from Case Western Reserve University in Cleveland, Ohio, and completed his residency requirements at the University of Rochester, UCLA, and King/Drew (Martin Luther King Jr./Charles R. Drew Medical Center) in Los Angeles. However, his experience as a black child in the segregated South seemed to shape his life even outside the region. At Case Western, Satcher become known as the "dignified activist" because of his push for and accomplishment of increased African American recruitment and enrollment at the university. He also began some community efforts in Cleveland to help underprivileged patients understand and use hospital services. During his time in California, he negotiated an agreement with UCLA School of Medicine and the Board of Regents that led to the establishment of a medical education program at King/Drew.

As surgeon general, appointed by President Clinton for a 1998–2001 term, Satcher became a spokesperson on issues such as obesity, smoking, youth violence, oral health, sexual health, and suicide prevention. Throughout his tenure as surgeon general, Satcher worked to eliminate racial and ethnic disparities in health care. In a 2002 PBS *NewsHour* interview, he said of his efforts, "I believe that to the extent that we respond to the health needs of the most vulnerable, we actually do most to promote the health of the nation." From 2002 to 2004 Satcher served as director of Morehouse College's National Center for Primary Care and in 2004 as the school's interim president.

Surgeon general Regina Benjamin, appointed by President Barack Obama in 2009, is another Alabama native. Benjamin's tenure, as of 2011, appeared to be influenced by her preappointment involvement in rural health care for uninsured patients on Alabama's Gulf Coast. Benjamin grew up in Daphne and Fairhope, Ala., and, like most other southern surgeons general, received her medical training in the South. She attended Morehouse College School of Medicine, along with Xavier University, in New Orleans (1979). She received

Regina M. Benjamin, the 18th surgeon general of the United States (U.S. Department of Health and Human Services, Office of the Surgeon General, Washington, D.C.)

her M.D. degree in 1984 from the University of Alabama at Birmingham and a degree in business administration from Tulane University in New Orleans in 1991. She completed a residency in family medicine in Macon, Ga., in 1987, and, per her agreement with the National Health Service Corps, which paid her tuition in exchange for a promise that she would work a specified number of years in an area with few doctors, she moved back home to the Gulf Coast to work. In 1990, she established a primary care clinic for mostly uninsured patients in the fishing village of Bayou La Batre, Ala. The clinic is still in operation, despite its being destroyed by two hurricanes (Georges in 1998 and Katrina in 2005) and a fire in 2006. Benjamin is still its CEO.

Her appointment as surgeon general drew some criticism from many people, who claimed that she, being somewhat overweight, was not an appropriate example of good health for the country. Those who championed her appointment said that Benjamin was merely able to relate effectively to the obesity problem plaguing the country. Otis Brawley, chief medical officer for the American Cancer Society, linked Benjamin's understanding of obesity to her background of work with impoverished people in Bayou La Batre: "Dr. Benjamin may understand the root causes [of obesity] and [be able to] effectively address the problems" from which it stems. Benjamin's understandings also reach to minority representation in health-care services. Before her appointment as surgeon general, she served on the Sullivan Commission, a group that studied diversity in the health-care work force. A report from the commission stated that, though the racial minority represents 25 percent of the country's population, minority physicians are only 6 percent—the same numbers found in a similar report from 1910. Benjamin hopes to help increase minority enrollment to overcome "ethnic and racial disconnect" between caregivers and patients.

Much of Benjamin's tenure so far has been focused on preventive medicine. The controversial 2010 health-care reform law created a commission to focus on a national government preventive medicine strategy and named Benjamin as its chairperson. In 1995 Benjamin was the first African American woman, and first physician under age 40, elected to the American Medical Association Board of Trustees. Benjamin also served as president of the State of Alabama Medical Association 2002–3. She was the country's first African American female president of a state medical society. Benjamin has also been the associate dean for rural health at the University of South Alabama's College of Medicine, and she was a 2008 recipient of a MacArthur Fellowship, which awards creativity in work, for her service to the diverse, underserved community in Bayou La Batre.

MARY AMELIA TAYLOR
University of Mississippi

Daniel Akst, *American Heritage* (December 2000); Joycelyn Elders, *Joycelyn Elders, M.D.: From Sharecropper's Daughter to Surgeon General of the United States of America* (1997); Amy L. Fairchild, *Public Health Chronicles* (May–June 2004); Ed Finkel, *Modern Healthcare* (1 March 2004); Frank R. Freemon, *Southern Medical Journal* (May 1987); Walter Hines Page and Arthur Wilson Page, eds., *The World's Work: A History of Our Time* (1912); P. N. Purcell and R. P. Hummel Jr., *American Journal of Surgery* (October 1992); Susan M. Reverby, *Tuskegee's Truths: Rethinking the Tuskegee Syphilis Study* (2000); E. Robert Wiese, *Southern Medical Journal* (October 1930).

Technological Education

Along with the rest of the nation, the South has long been aware of technology and its impact. Automobiles, computers, and hydroelectric power represent three of the most obvious examples of revolutionary change brought about by technology. Only in recent decades, however, has southern technological education kept pace with these changes or evidenced a high level of achievement.

Southern higher education before and after the Civil War was little different from that in the North. Focusing on the classical curriculum, a university education prepared the student for life as a cultured gentleman. Training in the sciences and practical fields such as engineering or agriculture had no place in the university experience. As Reconstruction ended, however, a growing awareness of the region's educational shortcomings emerged. When the American

Association for the Advancement of Science met in Nashville in 1877, the local press emphasized the value of science and technology to the region's economic well-being. Editorials stressed the need for improved educational facilities to provide young southerners with the tools needed to prosper in the new, more technologically focused economy of the New South.

During the 1880s, therefore, the recovering South began to reevaluate the role of higher education. Classical coursework was supplanted by curricula designed to prepare the student for a place in the world of business, industry, and agriculture. Many southern states established schools to supply such education. Auburn (Alabama Polytechnic from 1899 until 1960), Georgia Tech (founded as Georgia School of Technology in 1885), Clemson (1889), Texas A&M (1876), and Virginia Polytechnic Institute (1872) all owed their existence or continuation to the "practical" mentality of state legislators and businessmen, as well as to federal support through the Morrill Act.

The cultural and intellectual change represented by these new schools, however, may have been more apparent than real. As historian Thomas D. Clark argued in 1965, southern agricultural and mechanical colleges were chiefly concerned with improving the farming community's way of life. By focusing on land policy and management, as well as agricultural and general engineering, these institutions represented a commitment to the region's agrarian society. Another example of cultural inertia in education may be found in the development of separate schools for African Americans during the late 19th and early 20th centuries. Both northern philanthropists (guided by the recommendations of George F. Peabody) and state legislators supported the idea of "industrial education" for blacks. Yet this education, even at George Washington Carver's famous experiment station at Tuskegee Institute (1881), was designed to train efficient and contented laborers for the semi-industrialized southern agricultural economy.

The first few decades of the 20th century witnessed only slight improvement in southern technological education. The number of institutions providing appropriate instruction grew to include Louisiana Polytechnic Institute, Tennessee Polytechnic Institute, Texas Technological College, and others, but few major improvements in curricula or equipment took place. The Great Depression hit state-supported institutions especially hard and further thwarted the growth of such schools.

The end of World War II signaled the beginning of a major change in technological education throughout the United States. The war had been a total technological endeavor, symbolized by the development and use of the atomic bomb. Students entering college after the war (with or without the GI Bill) naturally found engineering and science courses attractive. The war experience had also emphasized the practical value of government support for research and development. Defense-related research, whether performed in government or university laboratories, enjoyed generous federal funding in the early years of the Cold War. Because of their relative lack of expertise, however, southern schools received little of this largesse, preventing significant growth of the region's technological education base. Although large sums flowed from Washington to such southern outposts as Redstone Arsenal (Huntsville, Ala.) and Oak Ridge, Tenn., southern education received few of the benefits.

The national shock of the successful launches of *Sputnik I* and *Sputnik II* in late 1957 precipitated another, and by far the most significant, change in America's educational history. For the next decade the nation focused on improving the scientific and technological base of American culture, with education receiving great attention and massive federal funding. Here, at last, the South began to enter the mainstream of technological and scientific development. The need for improved science education became immediately apparent, with southern schools at every level receiving federal funds to establish new programs and to improve existing ones. Recognizing another important regional need, the American Society for Engineering Education established its Historically Black Engineering Colleges Committee in 1964. Funded by such corporate donors as Exxon, IBM, General Electric, and others, this committee embarked on projects to improve the facilities and faculties at several such schools in the South. Aerospace research facilities, frequently cooperating with local universities, grew rapidly in Georgia, Florida, Alabama, Texas, and Louisiana. The University of Tennessee Space Institute was established in 1964, linking a graduate study program with research at the U.S. Air Force's Arnold Engineering Development Center. Schools such as the University of Alabama at Huntsville, Florida Institute of Technology, and Rice University worked closely with the National Aeronautics and Space Administration to guide the Apollo program to its successful lunar landing in 1969.

The post-Apollo letdown, however, witnessed a dramatic decline in both funding for and interest in science and technology. The aerospace industry in the South, which had been the region's major path to its share in the technological renaissance of the Space Age, withered perceptibly, removing employment opportunities in many technical fields and placing the region's technological education in a precarious position.

Within a few years, especially as "Sunbelt" industries demanded technically proficient workers, southern schools again found themselves stressing technological education. With high salaries available to scientists and engineers with undergraduate degrees, universities found themselves deluged with applications. In the South the region's population growth further reinforced this trend, as shown by the 1980 establishment of Southern Polytechnic State University (originally Southern Technical Institute) in metro Atlanta as the 14th senior college in the Georgia University System.

During the last two decades of the 20th century, the American technological landscape changed in several ways, creating new challenges and opportunities for education. Concerns about American international competitiveness were reinforced by Japan's growing role in the world economy, made possible by that nation's emergence as a leader in consumer electronics. The 1983 publication of *A Nation at Risk: The Imperative for Educational Reform* focused public attention on weaknesses in American public education, emphasizing that schools were not meeting the need for a competitive work force. Proposals for improvements in technological education were thus usually framed in terms of economic benefit, a perspective of special interest in the South as foreign automakers began to build assembly plants in the region. Although generous state and local incentives and a nonunionized labor force played important roles in attracting these facilities, the existence of a postsecondary educational infrastructure was often used to recruit manufacturers. Automakers were soon established in Alabama (Honda, Mercedes-Benz), Kentucky (Toyota), South Carolina (BMW), and Tennessee (Nissan, Volkswagen). Dell Computer Corporation also found the South an attractive region for expansion. The Texas-based company established manufacturing facilities in Tennessee and North Carolina in 1999 and 2004, respectively, providing yet another technology-oriented addition to the region's economy.

State education systems recognized and responded to the new demands, usually focusing on their community and junior colleges as the source of the required instruction. These schools had the added advantage of employing part-time instructors at lower salaries and minimum benefits, relieving stress on state budgets. The Mississippi State Board for Community and Junior Colleges, for example, not only oversaw the state's community college system but also had responsibility for the separate career and technical education and workforce education divisions, both of which focused on technological education. Southern states also recognized the importance of a well-coordinated effort to integrate their technological education facilities with continued recruitment of high-tech economic ventures. In 1983 Arkansas established its Science and Technology Authority to encourage the state's research and development activities, business innovation, and education in the fields of science, mathematics, and engineering. The same year also witnessed the creation of the Tennessee Technology Corridor in the eastern part of the state. The agency to coordinate this activity hoped to draw high-tech businesses into the region by emphasizing existing facilities such as Oak Ridge National Laboratory and the University of Tennessee, both of which had well-established educational programs. Even more immediately accessible was the State Technical Institute at Knoxville, whose name was changed in 1988 to Pellissippi State Technical Community College, precipitating dramatic enrollment growth during the next two decades.

Because of its status as one of the premier science and technology facilities in the nation, Oak Ridge National Laboratory played a major role in the South's educational efforts. In addition to various partnerships with the region's universities, Oak Ridge pursued an extensive outreach program to improve science education at the precollege level. Such endeavors increased after the University of Tennessee joined Battelle as the managing contractor for Oak Ridge National Laboratory in 2000. Grants to Tennessee high schools upgraded science classrooms and laboratories in an effort to increase the learning opportunities for students interested in science and technology.

Indeed, as the 21st century dawned, increasing emphasis on improving science and technology education at the K–12 level characterized much educational reform discussion. Continued poor performance of American students in assessments of science and technology knowledge suggested a serious potential decline in American competitiveness and led to a growing interest in STEM (Science, Technology, Engineering, Mathematics) education at the state level. In the South, STEM centers often appeared on university campuses, geared toward improving public school teachers' instructional efforts. North Carolina established its Science, Mathematics, and Technology Education Center in 2002 to coordinate educational efforts throughout the state and to work with educational, industrial, and governmental agencies toward this end. The Arkansas Science and Technology Authority, founded earlier to encourage technological development in the state, established a STEM Coalition to provide support and funding for improvements in such education at the K–12 level. With NASA funding, the Alabama Mathematics, Science, and Technology Education Coalition drew leaders from business, education, and government in its effort to improve the state's science and technology education. Throughout the region, STEM activities became increasingly visible, often securing external funding from local businesses and government.

At every educational level, the focus on technological education was increasingly tied to economic concerns. In late 2008, despite a severe budget shortfall that forced massive cuts to higher education, the state of Tennessee provided more than $11 million for technological education programs and facilities to support a new billion-dollar semiconductor plant in Clarksville. Few observers questioned the importance of high-tech industries to the South's economy. The support of technological education to attract such industries similarly brought few complaints. In the face of dramatic budget cuts to other educational programs, however, the South's emphasis on technological education raised troubling questions concerning the region's commitment to a well-educated citizenry, rather than a well-trained work force.

GEORGE E. WEBB
Tennessee Technological University

James D. Anderson, *History of Education Quarterly* (Winter 1978); Allan M. Cartter, *Southern Economic Journal* (July 1965); Thomas D. Clark, *Three Paths to the Modern South: Education, Agriculture, and Conservation* (1965); Eric Foner, *Reconstruction: America's Unfinished Revolution, 1863–1877*; Claudia Goldin and Lawrence R. Katz, *The Race between Education and Technology* (2008); Lawrence P. Grayson, *Engineering Education* (December 1977); Daniel S. Greenberg, *The Politics of Pure Science* (1967); Howard N. Rabinowitz, *The First New South, 1865–1920* (1992); *Rising above the Gathering Storm: Energizing and Employing America for a Brighter Economic Future* (2005); C. Vann Woodward, *Origins of the New South, 1877–1913* (1951).

Technology

Southern industry has traditionally included the processing of lumber, coal, and agricultural commodities. Such enterprises tended to perpetuate low wages and minimal skills. In fact, the agrarian tradition encouraged movement of the work force in and out of these industries on a seasonal basis. Early societal patterns seemed little affected by technology, although Eli Whitney's invention of the cotton gin in 1793 provided a technological foundation for the South's development. Most white southerners could easily identify with the position of ardent agriculturalists like Edmund Ruffin, a staunch advocate of the superiority of southern agrarian society in the antebellum era. Rural life was generally believed to be the most wholesome, moral, and virtuous form of existence. At the same time, technological enterprises like the iron industry, dating from the early colonial era, slowly advanced with the western and southern frontiers; by 1860 furnaces, forges, and rolling mills could be found from Delaware south to Georgia and as far west as Texas. Although the Civil War stimulated the growth of iron production in the South, even that industry lay in ruins by 1865.

In the late 19th century many farsighted southern leaders argued that the region needed to industrialize or forever remain a backwater of rural poverty.

The agrarian past would not be totally rejected; rather, a New South would have diversified, multicrop agriculture along with diversified manufacturing, lively commerce, and busy citizens. The best-remembered spokesman for the bold new departure was Henry W. Grady, editor of the *Atlanta Constitution*. Grady had watched a revitalized Atlanta emerge from the rubble of the Civil War. Commerce, industry, and urbanization, having worked wonders in Atlanta, could do the same for the Deep South. "The Old South rested everything on slavery and agriculture, unconscious that these could neither give nor maintain healthy growth," he argued in his famous New York address of 1886. The New South, he continued, represented a healthy democracy, "a hundred farms for every plantation, fifty homes for every palace—and a diversified industry that meets the complex need of this complex age." If the realities fell short of this generous ideal, and if overt racism and a patrician style of government still persisted, Grady's vigorous acceptance of urbanism and technology represented a significant shift away from the traditional patterns of culture.

The South's iron industry slowly recovered in the 1880s, especially in areas around Chattanooga, Tenn., and Birmingham, Ala. During World War I and World War II, modern techniques for steel production developed, resulting in new plants in Texas; in the postwar era, smaller, specialized facilities appeared throughout the South.

Factories and cities became more numerous in the South in the early 20th century, growing even more visible as a result of World War I and its urgent production requirements. Living patterns consistent with a technological society became more commonplace. During World War I and through the mid-1920s, lumbering made a considerable impact. Without completely disrupting rural social patterns, lumbering brought weekly paydays and tended to lessen the uncertainties of sharecropping.

When the momentum of the timber industry declined, southern sawmill workers did not return to the farm but moved to newly resurgent southern cities or industrial centers in the North. In order to retain year-round farm labor, farmers in industrial areas were forced to enter into new arrangements. Wages had to come closer to industrial levels, and a day's work came to mean 8 hours, not 16. In many older farming areas, technological society, as represented by industry, encountered stubborn hostility.

The process of industrialization became the catalyst for another technological phenomenon, the automobile. For years, the lack of adequate highways was seen as a serious shortcoming in attracting new industries and expanding existing ones. Just before World War I surfaced roads were so rare that textbooks in southern elementary schools included pictures of them as wonderful examples of the future. Aggressive highway commissions flourished in the 1920s, with gasoline taxes providing necessary revenues for road construction. Passable in every season, surfaced roads permitted large and small industries to spread throughout the rural South where railroads and rivers were nonexistent. For farmers, trucking brought new economic possibilities in marketing crops and livestock. After 1945 highways carried a flood of northern tourists in search of shrewdly marketed southern charm and a frost-free climate. Collectively, roads, cars, and trucks have helped end rural isolation.

The automobile shielded individual poverty from the public eye. On foot or astride a mule, poverty could be seen in ragged clothes and bare feet, but a car offered a technological cloak. Automobiles provided mobility, opening new horizons of change and opportunity. For poor southerners, the automobile became as significant as medicine or clothing. After World War II, southerners bought more cars than any other regional market group in the United States. Autos were an expression of individuality, as the prewar novels of William Faulkner and Robert Penn Warren show. In the postwar era, the role of the auto as a popular icon was evident in the huge throngs attracted to stock car races in places such as Darlington, S.C., and Daytona, Fla.

During the 1920s, as the South followed the seemingly irresistible patterns of commerce and industry "up North," there were still dissenters. The most celebrated example was the Agrarian movement at Vanderbilt University in Nashville, Tenn. Although the Agrarian critique was rooted in southern values and directed toward the southern scene, literary critic Louis D. Rubin Jr. has noted that it was consonant with other contemporary attacks on the materialism and depersonalization of 20th-century industrial society. In general, the Agrarians appealed to the younger generation to resist the onslaught of modern technology and harked back to an earlier era of southern agriculture as more harmonious and reasonable. Their manifesto, *I'll Take My Stand: The South and the Agrarian Tradition* (1930), seems not to have been taken seriously as an antitechnological tract by the great majority of southerners. In truth, the Agrarians did not intend to do away with technology but wanted to keep it within the bounds of a humanistic society. As Stark Young, a contributor to *I'll Take My Stand*, asserted, "We can accept the machine, but create our own attitude toward it."

In some respects, the Great Depression of the 1930s was an interlude, a time when industrialization, city building, and technological changes subsided, but only on the surface. The advent of highways and automobiles, the creation of the Tennessee Valley Authority, the spread of rural electrification, and the onset of World War II set in motion a series of changes of fundamental significance.

In the postwar era, inexpensive electricity supported the spread of air-conditioning to rural homes and urban offices alike, and the technology of the military-industrial complex left a pervasive imprint on southern culture. Army, navy, and air force installations dotted the region; Maryland, Georgia, and Texas became major centers of aerospace research and development and of manufacturing. As the nation's space program accelerated during the 1960s, the South achieved international attention owing to the location of several key installations of the National Aeronautics and Space Administration in Texas, Alabama, and Florida—a new "fertile crescent." The wartime stimulus of petroleum production and the refining industry was followed by increasing sophistication of the petrochemical industry, electronics, and medical research. These and other commercial-industrial trends have profoundly influenced educational patterns and career choices within an urbanized, industrial society.

The South's fierce attachment to the soil has been altered by an array of interrelated technological factors. Mechanization has changed the pastoral rhythms set by mules and plow horses; ancient landmarks such as boulder-strewn hummocks and fern-lined gullies have been leveled and filled for maximum farm production or for shopping malls. Although postwar industries have often provided the principal income necessary to allow rural families to stay on the "old place," while feeding a few cows and tilling a few acres on the side, much of the rural population has moved into burgeoning urban centers like Atlanta and Houston or to northern cities. More than ever, regional economics became tied to global vagaries of prices for oil, steel, pulpwood, and other commodities.

The spread of technology has contributed to some notable shifts in traditional cultural patterns. Attuned to the realities of the technological world, southern governors in the late 20th century banded together to urge Congress to raise certain tariffs to protect regional industry, a position that would have outraged southern Populists in the 1890s. The new realities also prompted southern civic leaders to renew criticism of the Ku Klux Klan and other racist groups because such organizations discouraged new industry from moving south.

Since 1950 the South's rate of economic expansion has been greater than the national average. New growth sectors include agribusiness, automobile manufacturing plants (28 percent of all automotive parts manufacturers and 31 percent of the nation's assembly capacity, as of 2011, is located in the South), defense industries, energy resources (oil and nuclear as well as coal, gas, and water), and high-tech research and development complexes such as North Carolina's Research Triangle Park. Educational and public services, along with race relations, have improved dramatically; but much industrial expansion is still dependent on a repressive, nonunion labor environment. (Research complexes such as North Carolina's Triangle Park encourage highly paid managerial and research personnel to migrate south without their unionized blue-collar work forces.) Average industrial wages in the region remain substandard. Most recently, globalization has disrupted traditional southern patterns, leading to the virtual collapse of the apparel and textile industries in the region, changes in the labor force because of transnational immigration, and new opportunities for major southern corporations.

ROGER E. BILSTEIN
University of Houston at Clear Lake City
DWIGHT B. BILLINGS
University of Kentucky

Richard M. Bernard and Bradley R. Rice, eds., *Sunbelt Cities: Politics and Growth since World War II* (1983); Blaine A. Brownell and David R. Goldfield, eds., *The City in Southern History: The Growth of Urban Civilization in the South* (1977); David L. Carlton and Peter A. Coclanis, *The South, the Nation, and the World: Perspectives on Southern Economic Development* (2003); Thomas D. Clark, *The Emerging South* (1961); James C. Cobb, *The Selling of the South: The Southern Crusade for Industrial Development, 1936–1990* (1993); James C. Cobb and William Stueck, eds., *Globalization and the American South* (2005); Louis D. Rubin Jr., ed., *The American South: Portrait of a Culture* (1980); Loyd S. Swenson, *Southwestern Historical Quarterly* (January 1968).

Urban Health Conditions

The southern region of the United States has historically been characterized by a poor health status and unfavorable ratings on most indicators of health and illness. This phenomenon can be traced back to pre–Civil War days when much of the region was isolated from the more "advanced" regions of the country. Not only did the region suffer from conditions that were not conducive to good health (e.g., poor nutrition, environmental hazards), but there was a general dearth of health-care resources to address the health problems that were so ubiquitous. The lack of resources was reflected in the absence of public sanitation programs and relatively fewer health personnel and facilities.

Early on, the unfavorable health status of the region was generally ascribed to the rural nature of the population. Southern states remain among the most rural and continue to be characterized by poor health status relative to the rest of the nation. Although the South has undergone a dramatic transformation since World War II, most of the historical health disparities remain. The southern states tend to rank higher than average on most indicators of health status. This is true in terms of the incidence of disease (e.g., diabetes, sexually transmitted infections), mortality rates (e.g., heart disease and stroke), infant and child mortality, and disability, as well as indicators of poor health such as smoking and obesity.

The diversification of the population of the region is reflected in the variations observed among the southern states in terms of health status. Although the states of the Deep South continue to rank poorly on most health indicators, Florida and Virginia tend to exhibit health indicators more in keeping with national trends. More than anything else, perhaps, this reflects the degree of urbanization in Florida and the growing urbanization of Virginia in the Washington, D.C., area.

Although the South's disadvantageous status in terms of health has persisted for decades, there has been a shift in the concentration of health problems over time. The health problems of the past could be attributed to the rural nature of the South, but today, with the region highly urbanized, it is found that the dominant health problems have shifted significantly from concentrations among rural populations to concentrations among urban populations. It is true that certain conditions remain higher among nonurban populations (e.g., child death rates, motor vehicle death rates), but these are typically more a function of the rural nature of those areas rather than anything distinctly southern.

When the overall health status of the nation's cities is considered, southern cities do not fare well. Sperling's Best Places index ranks the largest U.S. cities in terms of their health status based on five different dimensions. Although some might not consider this the most scientific approach to determining comparative health status, it does appear to be a reasonable indicator in terms of the claims made. This index indicates, for example, that only three southern cities (Austin, Raleigh-Durham, and Nashville) are included among the nation's top 25 healthiest cities. On the other hand, 13 southern cities are included among the nation's least healthy cities. New Orleans and San Antonio were identified as the least healthy, and most of the major metropolitan areas in the South, including Atlanta, Dallas, Houston, and Miami, made the least-healthy list.

A 2007 study carried out by *Forbes* indicated that southern cities accounted for 5 of the 10 most obese U.S. cities, and 10 of the 20 most obese. Among the leading offenders in this regard were Memphis, Birmingham, and San Antonio. The same study conducted by *Forbes* indicated high rates for one of the precursors of ill health—a sedentary lifestyle—for many southern cities. Southern cities accounted for 8 of the top 10 U.S. cities in terms of being sedentary and 12 of the top 20 U.S. on this indicator.

These overall figures, however, mask the true differences that exist in health status among southern cities. A major distinction exists between the health status of the urban cores of most of these cities and that of the suburban areas. While an argument can be made for the continued poor health of the South's rural areas, the region's inner cities follow close behind, leaving the South's suburbs as the one beacon of good health. Indeed, some of the region's inner cities rank among the worst in the nation in health status. Memphis, Tenn., is perhaps reflective of the situation within the region. Tennessee historically ranks among the bottom five states in health status. Shelby County, Tenn., typically ranks among the worst of the state's 95 counties in terms of health status. Within Shelby County, the city of Memphis has much worse health indicators than the surrounding communities and unincorporated areas. Within the city of Memphis the inner city has health-status indicators much worse than the state and the nation, thereby bringing down the overall rating for the county. On certain indicators, such as infant mortality, sexually transmitted infections, and other infectious diseases, Memphis's inner city is comparable to many Third World countries.

The poor health status of the South's inner cities tends to be correlated with concentrations of low-income minority populations. In much of the South, African Americans are concentrated within the inner city, frequently living in inferior housing and isolated from the majority population (and from access to health services). This is typical of cities such as Memphis and Atlanta. (New Orleans certainly epitomized this pattern of population distribution until Hurricane Katrina disrupted traditional residential patterns.) In other cases, there may be high concentrations of Latinos within the inner cities in some southern states, with Dallas and Houston being prime examples. And a polyglot city such as Miami exhibits concentrations of a wide range of ethnic groups within its inner-city area.

Although the poor health status of the South's urban areas is statistically associated with concentrations of minority groups, epidemiologists insist that the underlying contributor to poor health status is not race and ethnicity but socioeconomic status. Outside of the older metropolitan areas in the Midwest and Northeast, some of the nation's poorest populations are found in southern inner cities. In many cases, these concentrations of poverty have persisted over generations, with some being traced back to World War II–era rural-to-urban migration. Poor health status becomes perpetuated from generation to generation as a result of unhealthy lifestyles, unhealthy environments, and the impact of isolation and discrimination.

Poverty and its implications for health status are often exacerbated within southern cities as a result of a lack of access to health services. Certain southern cities are renowned for their health-care systems, but these systems frequently do not translate into adequate care for the sickest within these communities. Indeed, some of the worst conditions in the nation exist within the very shadow of major southern medical centers.

RICHARD K. THOMAS
University of Tennessee Health Science Center

Priscilla W. Ramsey and L. Lee Glenn, *Southern Medical Journal* (13 August 2002); Shakaib U. Rehman and Florence N. Hutchison, *Free Library* (16 June 2006); E. J. Roccella and C. Lenfant, *Clinical Cardiology* (December 1989); Rebecca Ruiz, *Forbes* (26 November 2007); Keith Wailoo, *Dying in the City of the Blues: Sickle Cell Anemia and the Politics of Race* (2001).

Barnard, Frederick A. P.

(1809–1889) EDUCATOR AND SCIENTIST.

Born in Sheffield, Mass., Frederick Augustus Porter Barnard spent half of his professional life in the South. Barnard received his A.B. degree from Yale in 1828 and, suffering from increasing deafness, he taught mathematics and geography at institutions for the deaf from 1831 to 1837. In 1838 he accepted a teaching position at the University of Alabama, where he hoped to pursue his developing interest in the sciences, especially astronomy, and higher mathematics.

Barnard taught mathematics, natural philosophy, and chemistry at Alabama. He was instrumental in the establishment of an observatory at the university, although he had to struggle with the board of trustees, the administration, and the state government for the funds. While in Alabama, Barnard cultivated an interest in early photographic techniques. He learned daguerreotypy from Samuel F. B. Morse and opened a gallery in Tuscaloosa in 1841. He maintained a scientific interest in photography and published articles in photographic journals throughout his career. During the years 1853–54 Barnard opposed a reorganization of the Alabama curriculum that would have implemented the same type of broad elective system used at the University of Virginia and Brown University. Barnard favored retaining a traditional discipline-oriented system in which the student would be allowed a few elective courses. The board of trustees, as well as Basil Manly, the president of the university, supported the Virginia plan and it was adopted. Barnard refused to work under the new system and resigned in 1854.

Barnard then went to the University of Mississippi, which had opened in 1848. The university badly needed instructors, and Barnard taught courses in mathematics, chemistry, physics, civil engineering, and astronomy. An ordained Episcopal minister, he also accepted a job as rector of the Oxford church. In 1856, only two years after his arrival, Barnard became president of the university. As president, his work was hampered by local residents, who saw no practical purpose for the university and regarded it with varying degrees of suspicion and dislike. Barnard was somewhat successful in his efforts to improve instruction and to acquire more sophisticated equipment for the school. Barnard's main interest at the University of Mississippi, as at Alabama, was the sciences, and his critics charged that he emphasized the study of science too heavily, to the exclusion of more traditional studies. As sectional differences intensified, suspicion of Barnard's northern roots pursued him. In March 1860 he was tried by the board of trustees on the charge of being "unsound on the slavery question." His supporters rallied behind him, however, and Barnard was cleared of the charges. Although Barnard was increasingly unhappy with his situation and attempted several times to secure positions elsewhere, poverty kept him from leaving Mississippi.

After Mississippi seceded from the Union early in 1861, university business was interrupted by the enlistment of many students in the military. Barnard left Mississippi in late 1861 and lived in Norfolk, Va., which Union troops captured in May 1862. Confederate president Jefferson Davis offered to hire Barnard to conduct a survey of the natural resources of the Confederacy, but Barnard refused because of his Union sympathies.

After working for two years with the U.S. Coastal Survey, Barnard was elected president of Columbia College, now Columbia University. During his 25 years at Columbia (1864–89), Barnard instituted standard entrance exams, introduced the concept of elective courses, and strengthened and enlarged the graduate and professional schools. His interest in science continued, and he was instrumental in founding the National Academy of Sciences in 1863. Barnard College, an official women's college of Columbia University, was established after his death and named in his honor.

Although Barnard later said that his years in Mississippi were among his worst, he seems to have genuinely loved the South and its people. His influence at the southern universities where he worked was felt long after his departure, especially in the area of science, and his hopes for upgrading the quality of education at Alabama and Mississippi were eventually fulfilled.

KAREN M. MCDEARMAN
University of Mississippi

William J. Chute, *Damn Yankee! The First Career of Frederick A. P. Barnard* (1977); John Fulton, *Memoirs of Frederick A. P. Barnard* (1896).

Cancer Alley (Louisiana)

Cancer Alley lies along a 100-mile stretch of the Mississippi River between Baton Rouge and New Orleans. The area has become a central case study for some of the most controversial and pointed issues in the modern environmental movement. The "Cancer Alley" moniker arose in the 1980s with the emergence of the environmental justice movement linking minorities and poverty to increased environmental hazards. Louisiana has a long-standing history of attempting to attract industry and business through tax incentives and plenty of available, cheap land. Cancer Alley, reflecting the monumental success of that policy, holds more than 300 industrial plants, as well as accompanying waste facilities. Many of the major petrochemical companies built plants in the area, including Texaco, Occidental Chemical, Chevron, Dow, and DuPont. Blacks continue to live there, historically an area of plantations, which today is also overwhelmingly poor. Although the chemical and industrial companies build their plants with a promise to provide jobs to the surrounding communities, most local residents lack a high school diploma and fail to qualify.

Residents claim a high incidence of illnesses associated with their proximity to the chemicals. These illnesses range from various cancers like leukemia to stillbirths, miscarriages, birth defects, and a variety of respiratory difficulties. The high level of pollution of various types, and perceived government neglect, led many communities to take a stand against their neighboring corporations. For example, in the small, mostly black community of Convent, La., Shintech, a Japanese corporation, proposed a massive polyvinyl-chloride plant in 1997. Convent, suffering from already high levels of toxins, organized to oppose the facility. The community achieved a small victory, defeating the original Shintech proposal, although the company built a smaller facility nearby. Children of Cancer Alley are intensely vulnerable to chemical contamination. In Gonzalez, La., a small community with a population of about 18,000, doctors discovered four cases of a very rare cancer, rhabdomyosarcoma, among children. The national rate for children with the disease rests at one in a million.

Although residents attempt to link the chemicals to ill health, industrialists and critics of the communities' theories, on the other hand, develop research to prove that the area's cancer rate closely approximates the national average. Other critics blame Cancer Alley's illnesses on the lifestyles of the residents, pointing to smoking and diet habits. Overall, despite much evidence of resident illness, communities have achieved very limited success against the chemical companies in the area. Cancer Alley remains one of the most polluted, and poverty-stricken, areas in the United States.

ELIZABETH BLUM
Troy University

Craig Colten, *Transforming New Orleans and Its Environs* (2000); *Green* (Laura Dunn, director, 2000); Frank D. Groves, Patricia A. Andrews, et al., *Journal of the Louisiana State Medical Society* (1996); Barbara Koeppel, *Nation* (8 November 1999).

Carver, George Washington
(c. 1864–1943) SCIENTIST.
Born in the final days of slavery in southwest Missouri, George Washington Carver was raised by his former owners, left home before his teenage years, and wandered until he was almost 30 years old seeking the elusive goal of many black contemporaries—a good education. After a brief career as an art major at Simpson College, he entered Iowa State, where his impressive abilities in botany earned him an invitation to pursue postgraduate studies. He received his master's degree in agriculture in 1896 and immediately accepted the position of director of agricultural studies at Booker T. Washington's Tuskegee Institute in Macon County, Ala. Although he intended to stay only long enough to establish a viable program, Carver remained at Tuskegee until his death in 1943.

For the first 20 years there, he labored under the shadow of Washington, endeavoring to improve the conditions of the poor and often landless black farmers of the South. He tried to provide inexpensive alternatives to costly commercial products at the experiment station he founded. His research and varied extension activities placed him in the mainstream, and sometimes the forefront, of agricultural education, but his idea of small-scale technology based on available and renewable resources was increasingly out of tune with the current trends. His efforts therefore brought limited recog-

George Washington Carver, botanist and agricultural researcher (Frances Benjamin Johnston, photographer, Library of Congress [LC-J694-302], Washington, D.C.)

nition, but they did aid thousands of individuals who were struggling under crushing burdens of debt within the sharecropping and tenancy systems of southern agriculture.

Carver's international fame came after Washington's death in 1915, when much of Carver's most useful work was over. His renown resulted mainly from his symbolic importance to myriad causes and was largely based on his essentially unsuccessful attempts to find commercial uses for such southern crops as peanuts and sweet potatoes. The eccentricities of his personality and the romance of his life story provided good copy for the press, and he was adopted as an exemplar by numerous groups—some with contradictory goals. He represented both the beneficence of slavery and segregation and the ability of African Americans. He was also used by the peanut industry, various religious groups, and New South editors preaching agricultural diversification and industrialization.

Although a growing mythology accompanied his rise to prominence, Carver continued to play an important role in the South. His warm, compelling personality led to numerous friendships with southern whites and provided a liberalizing influence on some newspapermen and many students whom he met during his lectures at white colleges. He also remained an inspiration to southern blacks, as he was repeatedly hailed as one of Dixie's leading citizens.

The George Washington Carver Museum was established at Tuskegee in 1942 and continues as part of the Tuskegee Institute National Historic Site. In 1943 the George Washington Carver National Monument at Carver's birthplace in Missouri became the first national monument to be dedicated to an African American and the first to a nonpresident. Further national recognition for Carver came in 2000 when he became a charter inductee in the United States Department of Agriculture's Hall of Heroes, which named him the "Father of Chemurgy," and in 2005 the American Chemical Society designated his work at Tuskegee a National Historical Chemical Landmark.

Aside from national recognition in the form of monuments and historic sites, numerous museums and schools in the South bear Carver's name. Many African American public schools in the 1940s and 1950s were named after him, and, though many became casualties of desegregation, several schools remain, including George Washington Carver High Schools in Birmingham and Montgomery, Ala., and in Columbus, Ga. Many sites where Carver schools were once located have maintained their ties to his legacy through museums and venues, such as the Central-Carver High School Museum in Gadsden, Ala., and the Carver Theater for the Performing Arts in Birmingham, Ala. Dothan, Ala., nicknamed the "Peanut Capital of the World," pays homage to Carver's recognition of the lowly peanut, which made it an invaluable economic booster for the area, with the George Washington Carver Museum, Monument, and a National Peanut Festival. Though many of Carver's accomplishments have been overshadowed by advancements in technology, his name and legacy still continue as an enduring presence in the South.

LINDA O. MCMURRY
North Carolina State University

MARY AMELIA TAYLOR
University of Mississippi

George Washington Carver Papers, Tuskegee Institute Archives, Tuskegee, Ala.; Mark D. Hersey, *My Work Is That of Conservation: An Environmental Biography of George Washington Carver* (2011); Gary R. Kremer, ed., *George Washington Carver in His Own Words* (1987); Linda O. McMurry, *George Washington Carver: Scientist and Symbol* (1981), "George Washington Carver," *Encyclopedia of Alabama*, http://encyclopediaofalabama.org (2007); National Park Service, "George Washington Carver National Monument," http://nps.gov/wca (2010), "Tuskegee Institute National Historic Site," http://nps.gov/akr/tuin (2010).

Centers for Disease Control and Prevention

The Centers for Disease Control and Prevention (CDC), an Atlanta-based U.S. Federal Agency under the Department of Health and Human Resources, works to develop and implement disease control and prevention measures nationally, locally, and globally. It encourages disease-prevention policies in state and local governments and assesses and develops measures against national bioterrorism and environmental threats.

The CDC began in 1942 as the Malaria Control in War Areas (MCWA) agency, whose primary goal was the containment and eradication of malaria. Because the southern part of the United States was particularly susceptible to malaria outbreaks, and because much of the U.S. basic military training took place on southeastern military bases, Atlanta was chosen for the location of the MCWA headquarters. The MCWA's original staff of scientists, entomologists, and engineers studied the malaria disease and mosquitoes, the agents for its spreading. In its initial years, the center's primary goal was assisting states in the containment and eradication of that disease, and it implemented such strategies as the spraying of DDT in homes and neighborhoods. After World War II, a doctor named Joseph W. Mountin converted the MCWA into a peacetime agency called the Communicable Disease Center. The center's mission was expanded to the collection and analyzing of information about many infectious diseases, not just malaria, and to the dissemination of that information to public health officials. Atlanta remained the headquarters for the CDC, and Emory University provided land for the center's buildings in 1947.

In 1970 the CDC was renamed the Center for Disease Control because it had taken on numerous public health programs in addition to those that studied and controlled infectious diseases. New programs included those that dealt with nutrition issues, chronic diseases, and family planning. The National Institute for Occupational Safety also joined the CDC in 1970, further expanding the CDC's authority into the realm of environmental and workplace health. The subsequent years saw tremendous expansion of the CDC's work in the public health field. No longer confined to the control of communicable diseases, the CDC became increasingly involved in the control and treatment of chronic health problems, such as obesity, diabetes, autism, cancer, and STDs. The CDC's research also began to affect food safety standards and regulations for tobacco use. In 1992 the CDC kept its acronym but became the Centers for Disease Control and Prevention to reflect its ever-widening influence in the nation's public health. In 2005 the CDC added the Emerging Infectious Diseases Laboratory, a Bio-Safety Level 4 facility used to study extremely dangerous diseases that are countered by no effective vaccines, such as the Ebola and avian flu viruses. The CDC also added the Environmental Health Lab in order to develop prevention techniques for diseases caused by toxic chemicals in the environment and to effectively respond to bioterrorism threats and public health emergencies.

The CDC Foundation, a nonprofit organization created by the Public Health Service Act in 1994, directs the CDC's connection to the private sector, aiming to "[close] the gap between science and action." The CDC Foundation helps procure funding for more quickly and effectively accomplishing CDC tasks. Those tasks have increasingly reached both farther inward and farther outward, beginning programs specific to local communities in the United States and working for public health on a global scale. In addition to encouraging state and local governments to promote initiatives like increased physical activity in school physical education classes, the CDC has worked with local communities to develop strategies for communitywide health. CDC's Healthy Communities program helped Savannah, Ga., to establish a citywide ban on smoking in public buildings, including nightclubs and bars. The same program has, since 2004, worked to encourage primary care and education for people with chronic illnesses in Alabama's River Region in order to curtail superfluous emergency room visits. By 2007 emergency visits in the area had decreased by 50 percent.

Though the CDC had been officially globally active since 1958, when it first sent a team to Southeast Asia to respond to cholera and smallpox outbreaks, the agency has only increased its global involvement through the present. The CDC's Center for Global Health has implemented many programs to help alleviate global problems with AIDS, HIV, TB, cholera, malaria, polio, and influenza. The center also works to pro-

The Centers for Disease Control and Prevention's Roybal campus, a high-tech laboratory/vivarium (an enclosed area for keeping and raising animals or plants for observation or research), in Atlanta, Ga. (Photo by James Gathany, Centers for Disease Control and Prevention, PHIL ID #10693)

mote improved women's health and access to clean water and sanitation in other countries. In October 2010 the CDC began working closely with other U.S. and global health organizations to assist Haiti's Ministry of Public Health and Population in containing a cholera outbreak in Haiti, following a devastating earthquake there in January 2010. In addition to the CDC's extensive Web site, which provides limitless information on the agency's global influence, a new 2009 CDC–Emory University Web site, the Global Health Chronicles, has begun creating a digital history of global disease prevention aimed at educating people about global public health, including the CDC's role in it.

The CDC has become increasingly heavily involved in local, national, and global disease control and prevention. However, almost as much of the CDC's focus is on education. In addition to the dissemination of vital disease prevention information to state and local health officials, the agency established the Global Health Odyssey Museum at the Atlanta headquarters in 1996, the 50th anniversary of the CDC and the year of Atlanta's hosting of the Olympic Games. The museum, established in conjunction with the Smithsonian Institution, collects and displays historical materials and equipment, like early malaria-eradication equipment and an iron lung. The museum has since expanded to host many changing exhibits and annual programs, including the Disease Detective Camp, an educational program for high school juniors and seniors begun in 2005 to educate students in public health and CDC's actions to promote it. The agency remains a service for improving the public's health through education, policy reforms, and tangible assistance in the creation of effective vaccines and implementation of prevention techniques.

MARY AMELIA TAYLOR
University of Mississippi

Centers for Disease Control and Prevention, www.cdc.gov (2011); CDC Foundation, www.cdcfoundation.org (2011); Emory University, *The Global Health Chronicles*, www.globalhealthchronicles.org (2009); Elizabeth W. Etheridge, *Sentinel for Health: A History of the Centers for Disease Control* (1992); Bindu Tharian, "Centers for Disease Control and Prevention" and "Malaria," *New Georgia Encyclopedia*, www.georgia encyclopedia.org (2009).

Country Doctor

Over the years, the image of the country doctor has taken various forms. One is a romantic image of the southern country doctor who administers to the needs of his community. This scene is aptly illustrated in Luke Fildes's picture *The Doctor* (1891), in which the doctor leans over his young patient, a girl asleep on a row of chairs with her parents in the background, fretful and worried about her well-being. For others, a doctor's bag and stethoscope or a horse and buggy have become symbols of the country doctor. The reality of a rural medical practice and the true character of a southern physician surely comprised complex challenges and great personal reward.

Doctors practicing in the American South during the 19th century believed the practice of medicine in the South was distinctive to this area and thus different from other areas of the country, especially the northeastern states. Although many southern doctors attended medical schools located in northeastern cities, graduates were usually quick to return to their southern homes to establish their practices. During the antebellum period, country doctors were called to treat whites, free blacks, and slaves. The humid, warm climate of the South created an almost subtropical ecological zone where certain illnesses could flourish and sweep through a population. The African diseases of malaria, yellow fever, and hookworm were imported into the South along with the large population of enslaved Africans. These diseases became widespread and caused southern residents to suffer significant mortality rates, prolonged weakness, and increased vulnerability to other illnesses.

Against a backdrop of recurring illnesses and diseases, a southern country doctor faced his share of occupational challenges. Most rural patients could not pay cash when the physician rendered his services and would have an account with the doctor. The doctor might send an annual bill for charges incurred, but payment was uncertain at best; cash was preferred, but in-kind payments of chicken eggs, hogs, peaches, turkeys, hams, or other commodities were usually accepted to settle the bill. House calls to visit an ill patient might entail arduous travel over long distances in all forms of inclement weather. Ineffective medicines could frustrate the most dedicated physician.

Before the Civil War, southern doctors were white men with a few women in their ranks. Surgical training was still in its infancy and medical therapies dominated the practice of rural medicine. After the Civil War, country doctors treated both whites and blacks, while the South began to see a growing number of African American doctors. A country doctor relied on various resources to practice good medicine: formal training at medical schools or apprenticeships, staying current with new medical books and journals, contributing writings to the field, participating in medical societies, and traveling throughout the area or abroad to learn new techniques and methods. Even still, as in the image of *The Doctor*, sometimes all he could do was wait for the patient to take a turn, either for better or for worse.

Rising above these everyday obstacles was the country doctor's commitment to the medical field, his practice, community, and patients. Community oriented, a country doctor practiced with a personal knowledge of the individuals and families in his area. He knew not only a person's medical history but also his family upbringing, economic situation, and social standing. The physician's own standing in the community was often one of high regard, especially when the doctor also took an active role in community life.

After the 1880s doctors benefited from an expanded scientific knowledge base including germ theory, antiseptics, sterilization techniques, and more effective medicines. Migration of rural residents to the industrialized cities, as well as the increasing availability of the automobile and telephone, also affected the physician's practice. Urban physicians began to practice in more hospitals and clinics than their rural counterparts; this pressure for modern medical facilities and clinical settings helped contribute to the decline of house calls.

Even with significant changes in the medical field, the southern country doctor's dedication to the community he served remained. In the early 20th century, country doctors continued to deliver many home births, help organize vaccination clinics in their rural communities, and prescribe more powerful and effective medicines. In keeping with his community-focused practice, the country doctor himself became a steadfast symbol of rural American life.

ANNE ANDERSON
The Country Doctor Museum
Bailey, North Carolina

John Duffy, *From Humors to Medical Science: A History of American Medicine* (1993); Kay K. Moss, *Southern Folk Medicine, 1750–1820* (1999); Ronald L. Numbers and Todd L. Savitt, eds., *Science and Medicine in the Old South* (1989); Todd L. Savitt, *Fevers, Agues, and Cures: Medical Life in Old Virginia* (1990); Todd L. Savitt and James Harvey Young, eds., *Disease and Distinctiveness in the American South* (1988); Steven Stowe, *Doctoring the South: Southern Physicians and Everyday Medicine in the Mid-Nineteenth Century* (2004).

Creation Science

Creation science, or scientific creationism, holds that scientific evidence supports the biblical refutation of Darwinian evolution as an explanation for human existence. Usually identified with fundamentalist religion, creationism's proponents aim to prove the biblical account of creation using the scientific method. The creation-evolution battle has been fiercely waged since the 1920s and the highly publicized Scopes Trial in Dayton, Tenn., and much of the struggle has consistently been fought on southern soil. The South led the 1920s fight to remove the theory of evolution from public school curricula. At that time, four southern states—Arkansas, Florida, Mississippi, and Tennessee—passed legislation restricting the teaching of evolution, and the antievolutionists responsible for these laws (led by William Jennings Bryan) openly defended their views as biblically based.

Today's controversy over scientific creationism differs significantly from the 1920s battle and has undergone many changes, but the foundational arguments remain. In the 1970s the creationism movement was revived with a new focus, influenced by the efforts of Henry M. Morris, a Texas engineer. Morris, raised a Southern Baptist, argued in *The Genesis Flood* (1961) for a return to the belief in a literal six-day creation and a worldwide flood as described in the Bible. His strict creationism was hotly debated within the scientific community. In the 1970s Morris joined with fundamentalists outside the scientific world in a renewed effort to propagate creationist ideas. The new fight stressed the scientific aspect of creationism, offered evidence for a worldwide catastrophe, and focused on arguments against evolution. In 1970 Morris founded the Institute for Creation Research (ICR), a center devoted to the application of scientific research toward proving the biblical creation narrative and disproving evolution. The ICR moved from San Diego to Dallas, Tex., in 2007. In 1982 the U.S. Supreme Court ruled that creationism was not viable as a scientific theory, and the ruling removed the Arkansas law that required the equal treatment of creation science and evolution in the state's public school curriculum. The 1982 *McLean v. Arkansas* case joined a 1968 Supreme Court case, *Epperson v. Arkansas* (which struck down states' bans on the teaching of evolution) on the side of evolution science, and in 1987 a U.S. Supreme Court case from Louisiana, *Edwards v. Aguillard*, stopped the teaching of creation science in public schools altogether.

In response to these legal setbacks, in 1984 an offshoot of creationism appeared in hopes of gaining credibility among the scientific community and victories in court battles: intelligent design. In an attempt to restore creationism to public school classrooms while skirting complaints and court cases about advancing particular religious beliefs, a Texas-based Christian organization, the Foundation for Thought and Ethics, published a school-level textbook, *Of Pandas and People: The Central Question of Biological Origins*, introducing "intelligent design" as an alternative to evolution. This origins theory removes the mention of God or a specific religious belief and merely asserts that every life form is too complex to have happened by chance through Darwin's theory of natural selection. Rather than naming God, or a "creator," intelligent design proponents point to the existence of a "Designer" that orchestrates origin and evolutionary processes. In 2005 the U.S. Supreme Court ruled that intelligent design was a form of creationism and therefore also not allowed in public school curricula.

It seems that the American public is still divided about creation science and evolution, but southerners still seem to favor creationism over evolution, despite court battles and the chagrin of the National Academy of Sciences and other scientific institutions. In 1998 a Southern Focus Poll revealed that 67.7 percent of southerners claimed to believe the biblical creation account of the origins of life, whereas only 17.1 percent believed the Darwinian evolution theory. In nonsouthern states, belief in creationism was much less prevalent (50.6 percent), and Darwinian evolution was almost twice as popular in those states than in the South (33.5 percent). In 2009 the Harris Poll and BBC World News America concluded that "people in the South and people who describe themselves as religious . . . are less likely to believe that humans evolved from earlier species than are people in other regions or people who are not religious." Although the poll did not indicate what percentage of self-described "religious" subjects actually live in the South, the mention alone of the South and religiousness in the same sentence and with corresponding results is indicative of the pervasiveness of the influential culture of religiosity in the South.

The argument between religious fundamentalism and public education has repeatedly surfaced in the South, where conservatives and fundamentalists have organized and pressured local and state school districts and legislators. Though Supreme Court cases (all but one took place in southern states) have consistently ruled in favor of evolution science, the debate still continues over whether both theories are viable and whether both or only one should be taught in schools. In July 2010 the Livingston Parish, La., School Board meeting discussed its desire to reintroduce creationism to its classrooms. Immediately a flurry of protests began, and a few days later the board announced that no changes to curriculum would be made in 2010, but that the board would still consider reincorporating creationism in future curricula. Such action would be sure to draw down more heated debate, and as the 2009 Harris Poll indicates, "the argument is unlikely to end any time soon."

KAREN M. MCDEARMAN
MARY AMELIA TAYLOR
University of Mississippi

Melinda Beck, *Newsweek* (28 July 1986); William A. Dembski and Jonathan Wells, *The Design of Life: Discovering Signs of Intelligence in Biological Systems* (2007); Gary B. Ferngren, ed., *Science and Religion: A Historical Introduction* (2002); John Hill, *Jackson Clarion Ledger* (5 October 1986); Institute for Creation Research, www.icr.org (2010); Lee Ann Kahlor and Patricia A. Stout, eds., *Communicating Science: New*

Agendas in Communication (2010); J. B. Labov and B. Kline Pope, *CBE Life Science Education* (Spring 2008); Edward J. Larson, *Summer for the Gods: The Scopes Trial and America's Continuing Debate over Science and Religion* (1998), *Trial and Error: The American Controversy over Creation and Evolution* (2003); National Academy of Sciences, *Science and Creationism: A View from the National Academy of Sciences* (1999), *Science, Evolution, and Creationism* (2008); Ronald L. Numbers, *The Creationists: From Scientific Creationism to Intelligent Design* (2006), *God and Nature: Historical Essays on the Encounter between Christianity and Science* (1986); Andrew J. Petto and Laurie R. Godfrey, eds., *Scientists Confront Intelligent Design and Creationism* (2007); Eugenie C. Scott, *Evolution versus Creationism: An Introduction* (2004); Christopher Tourney, *God's Own Scientists: Creationists in a Secular World* (1997).

DDT

First developed by the Swiss chemist Paul Muller in 1939, DDT (from dichlorodiphenyltrichloroethane) has been one of the most widely used pesticides in the world. During World War II, DDT helped alleviate two major problems for U.S. forces stationed overseas, typhus and malaria. Much of the early research for wartime use was conducted at the Orlando, Fla., laboratory of the Bureau of Entomology and Plant Quarantine and as part of the Public Health Service's Office of Malaria Control in War Areas, which conducted DDT experiments throughout the South.

After the war, DDT became the predominant form of insect control in the United States, especially in the South. During the postwar years, DDT remained an important chemical for southern farmers despite the development of other chemical pesticides. By 1970 nearly 70 percent of all DDT used nationwide was sprayed on southern crops. The largest users were cotton growers, who relied on the chemical to control the historically troublesome boll weevil. To serve the southern market, between 1947 and 1970, the Olin Corporation leased the Redstone Arsenal in Huntsville, Ala., from the Army Corp of Engineers to produce DDT for the army and local companies. The plant produced millions of pounds of the chemical.

DDT is a broad-spectrum, persistent pesticide. It does not target one insect, but kills different species of insects. Once sprayed, DDT remains active for nearly six months. Its metabolite, DDE, however, can be found in the natural world years after the initial application. The chemical migrates and accumulates in the fatty tissues of all animals, including humans. Animals at the higher end of the food chain have higher concentrations since DDT is absorbed with digestion. Fish and fowl are equally susceptible to the chemical's toxicity.

Because of its importance to cotton production and the southern economy, some of the most vocal advocates for DDT came from the South. This was particularly noteworthy after the publication of Rachel Carson's *Silent Spring* (1962), a book that exposed the long-term hazards of DDT. Over the next decade, environmentalists fought to prohibit the use of DDT at state and national levels. Southern environmental activists and political leaders often took different sides of the debate.

Many southerners campaigned for environmental protections against DDT because of its ecological hazards. In 1962 an *Atlanta Journal-Constitution* headline read, "Insecticides . . . Big Poison Is Destroying Our Wildlife." Outdoorsman and columnist of the *Charlestown News-Courier* Roy Attaway, an outspoken critic of chemical pesticides, once commented, "It is not pleasant to realize that you and I and every citizen of the United States have lethal poisons stored in our tissue."

DDT is a known endocrine receptor, which can cause premature birth, miscarriage, congenital hypothyroidism, and other health problems in humans. It has been also linked to neurological disorders (such as Parkinson's) and asthma. Recent studies also indicate that women exposed to DDT at a young age have higher rates of breast cancer than the general population. Despite years of scientific research, we still know very little about DDT's long-term health impacts.

Southern congressional leaders condemned the environmental movement and its critique of chemical pesticides. Mississippi representative Jamie Whitten (D) and Texas congressman William R. Poage (D), who served as chair of the House Committee on Agriculture from 1967 to 1975, were two of the most outspoken critics of DDT regulation. Jamie Whitten, for his part, wrote a book-length response to *Silent Spring*, titled *That We May Live* (1966), which argued that pesticides like DDT were critically importantly to agricultural development. Similarly, William Poage introduced legislation to give the Department of Agriculture veto power over the regulatory authority of the Environmental Protection Agency.

The debate over DDT essentially came to an end in 1972 when the Environmental Protection Agency banned the pesticide for domestic use. The long-term impacts of DDT, however, continued to shape the history of the South in the decades to come.

In 1979, 1,200 residents of Triana, Ala., filed suit against Olin Corporation. The Tennessee Valley Authority reported Olin released nearly 4,000 tons of DDT into the Huntsville (Ala.) Spring Branch near the site of the Redstone Arsenal. Fish in the river contained over 200 parts per million of DDT, or nearly 40 times over the federal limit. Residents who ate fish from the river also had usually high levels of DDT, some as high as 3,300 parts per billion. In 1982 Olin agreed to pay damages and clean up the Huntsville Spring Branch. By 1995 the cleanup was completed.

DAVID KINKELA
SUNY *Fredonia*

Pete Daniel, *Toxic Drift: Pesticides and Health in the Post–World War II South* (2005); Thomas Dunlap, *DDT: Scientists, Citizens, and Public Policy* (1983); Ethel Hall, with Carmelita James Bivens, *My Journey: A Memoir of the First African American to Preside over the Alabama Board of Education* (2010); David Kinkela, *DDT and the American Century: Global Health, Environmental Politics, and the Pesticide That Changed the World* (2011); James McWilliams, *American Pests: The Losing War on Insects from Colonial Times to DDT* (2008); Edmund Russell, *War and Nature: Fighting Humans and Insects with Chemicals from World War I to "Silent Spring"* (2001); Robert Van Den Bosch, *The Pesticide Con-*

spiracy: An Alarming Look at Pest Control and the People Who Keep Us "Hooked" on Deadly Chemicals (1989).

DeBakey, Michael

(1908–2008) SURGEON.

Dr. Michael Ellis DeBakey, a longtime fixture in the Department of Surgery at the Baylor University College of Medicine in Houston, Tex., ranked among the world's leading authorities in the field of cardiovascular research and was a significant influence in transforming Houston into a major international medical center.

Michael DeBakey, pioneering heart surgeon, 1970s (Texas Collection, Baylor University, Waco, Tex.)

Born 7 September 1908 in Lake Charles, La., to a Lebanese immigrant family, DeBakey received his M.D. degree from Tulane University in 1932. He worked for some time under New Orleans surgeon Dr. Alton Ochsner and then served in the Surgical Consultants' Division in the Office of the U.S. Army Surgeon General during World War II. While working with the surgeon general's office, DeBakey designed the mobile army surgical hospital (MASH) units, which allowed doctors and surgeons, not just first-aid workers, to treat wounded soldiers at the front lines. The MASH units were first used in the Korean War, and, in addition to saving the lives of numerous soldiers in the Korean and Vietnam wars, they provided inspiration for the popular television series *M*A*S*H**. In 1945 DeBakey became director of the Surgical Consultants' Division and received the Legion of Merit for his service to the army. In 1948 he moved to Baylor College of Medicine as chair of its Department of Surgery and quickly began to earn laurels for himself and the school.

DeBakey's achievements in the world of medical science were many. At age 23, while still a medical student at Tulane, he invented the roller pump, which years later became a key component in the heart-lung machine and made open-heart surgery possible. He developed artificial grafts for cardiac bypass surgery and an artificial ventricle to be used on a temporary basis by heart surgery patients, and he participated in the attempt to design an artificial heart to replace the human organ. He was a pioneer in heart transplant surgery and was the first surgeon to successfully implant a partial artificial heart. A skilled surgeon, he operated successfully on more than 60,000 patients, only discontinuing his regular surgery schedule in his 80s. He published hundreds of scholarly articles to report on the results of his research and was granted many scientific awards and honorary degrees, both in the United States and abroad. In 1969 President Johnson awarded DeBakey the Presidential Medal of Freedom, and in 1987 President Reagan awarded him the National Medal of Science. In 1999 the United Nations honored him with its Lifetime Achievement Award, and his work on the DeBakey Ventricular Assist Device earned him the 2001 NASA Invention of the Year award. In April 2008, at age 99, DeBakey was awarded the Congressional Gold Medal, the highest civilian honor granted by Congress.

In addition to his scientific achievements, DeBakey also distinguished himself as an outspoken advocate of government support for various facets of American medicine. He led the movement to start the National Library of Medicine in the 1950s, and he was appointed by both presidents John F. Kennedy and Lyndon B. Johnson to serve on advisory councils that, despite the opposition of organized medicine, recommended that federal funds be used for regional programs to improve patient care for victims of heart disease, cancer, and stroke. He also stressed the need for government sponsorship of pure research and medical education.

During the nearly four decades of his association with Baylor, DeBakey served in various leadership capacities, and he was largely responsible for the development of two of the world's leading centers of heart surgery—the Methodist Hospital and the Texas Heart Institute, both located in Houston. DeBakey was also instrumental in establishing the Veterans' Affairs Medical Center in Houston, which, along with numerous divisions and programs at Baylor and the Methodist Hospital, now bears his name.

Dr. DeBakey died of natural causes on 11 July 2008, just a few months short of his 100th birthday, but his contributions to the field of cardiovascular medicine are enduring.

LUCIE R. BRIDGFORTH
Memphis State University

MARY AMELIA TAYLOR
University of Mississippi

Lawrence Altman, *New York Times* (25 October 2006; 1 May 2007; 13 July 2008); Michael E. DeBakey and Antonio M. Gotto, *The New Living Heart Diet* (1996), *The New Living Heart* (1997); Methodist DeBakey Heart and Vascular Center, www.methodisthealth.com (2010); *Saturday Review* (16 October 1971); *Time* (28 May 1965).

Faith Healing

Faith healing is deeply rooted in southern folklife, often connected to spirituality, as in African-derived practices, Native American lore, Latino *curanderismo*, and white folk customs tracing back to European magic traditions. The *Foxfire* project documents southern Appalachian healing practitioners, who treat burns, thrush (a childhood mouth ailment), and any health problem they see as not having natural causes. The recitation of verses from the Old Testament is often a part of the healing procedure. "I do all that by the help of the Lord," said one healer quoted by interviewers.

In the late 19th century, the Pentecostal-Holiness movement emerged, foregrounding healing as part of a "fourfold" gospel that also included personal salvation, Holy Ghost baptism, and the second coming of Christ. Divine healing goes far back in Christianity, but the Protestant-Holiness theology stressed that Christ's sacrifice on the cross offered healing for the body as well as the soul, an idea that had not been a traditional part of Christian belief. Pentecostals understood Christ's atonement as a causal agent in healing human bodies. A network of conferences, Bible institutes, colleges, publications, and local associations popularized the idea. Pentecostals came to insist the "prayer of faith" could automatically heal the body. They cited such biblical texts as the Great Commission in Mark 16:17–18, the gifts of healing in Corinthians 12:9, and the church elders healing responsibilities in James 5:14. As historian Grant Wacker notes of this tradition's belief, "Since God had promised to mend the body if one prayed with genuine faith, the only possible result was immediate and complete restoration of the body." Inevitable failure in some healing attempts did not distract healers from citing examples of seeming success, and early Pentecostals had thousands of testimonies of successful healings. These beliefs were often accompanied by suspicions of medical practitioners. At a time before full professionalization of medicine, early Pentecostals saw physicians as either incompetent quacks or greedy leeches.

Historian David Edwin Harrell identified a "healing revival" from 1947 to 1958, and it marked a flourishing of faith healing in the South. Baptist minister William Marrion Branham began the revival in Pentecostal churches in Arkansas and Missouri. Earlier revivalists who had used healing in their meetings early in the 20th century often reappeared in these new revivals. Among the leading healing revivalists were A. A. Allen, born in Oklahoma City and based in Dallas; Jack Coe, also working out of Dallas; and T. L. Osburne, centered in Tulsa but taking revivalistic healing around the world. The *Voice of Healing* was the periodical that published extensive healing testimonies.

Oral Roberts was the best-known faith healer of the 20th-century South. Born the son of a preacher in Oklahoma, Roberts had been healed of tuberculosis as a teenager and was ordained a minister shortly afterward in the Pentecostal-Holiness Church. His itinerant ministry included small-town revivals and pastoring small churches in the South and Southwest. On the basis of his biblical reading, Roberts in his early ministry nurtured the hope of healing gifts, and in 1947 he performed his first successful healing, becoming a higher profile evangelist after that. He moved to Tulsa in 1948, established Healing Waters, Inc. (a nonprofit ministry), and within five years was preaching to more than a million people a year as well as reaching a radio audience over 300 stations. He developed a television ministry in 1954, with more than 136 stations showing his big-tent services.

Roberts ended his healing revivals in 1968, becoming a United Methodist minister but retaining his interest in faith healing. He had become a student of psychosomatic illness and argued for the connections between the mind, soul, and body that affected divine healing. In the late 1960s his message was increasingly one of positive thinking and the possibilities of not only body healing but material abundance through faith and prayer. He had never been hostile to professional medicine, and in 1977 he revealed that a 900-foot Jesus had told him to combine medicine and faith through a multimillion-dollar City of Faith Medical and Research Center, with a hospital and medical school to be built on the campus of his Oral Roberts University in Tulsa. Roberts launched an ambitious fundraising effort, with an increasingly materialistic tone that drove away many of his earlier supporters. The City of Faith opened in 1981, but financial problems led to its closure in 1989.

The Pentecostal belief in faith healing rested in the confidence that the miracles seen in the Bible could still be performed in the modern world. In the 1980s Oral Roberts, along with his family, appeared on a new television show, *Miracles Now*, and in the 1990s his son Richard carried on the tradition hosting an evening television show, *The Hour of Healing*.

CHARLES REAGAN WILSON
University of Mississippi

Nancy Hardesty, *Faith Cure: Divine Healing in the Holiness and Pentecostal Movements* (2003); David Edwin Harrell, *All Things Are Possible: The Healing and Charismatic Revivals in Modern America* (1978); Robert T. Trotter II and Juan Antonio Chavira, *Curanderismo: Mexican American Folk Healing* (1981); Grant Wacker, *Heaven Below: Early Pentecostals and American Culture* (2001); Eliot Wigginton, ed., *The Foxfire Book* (1972).

Frontier Nursing Service

In the 1920s experts expressed alarm that the United States' maternal and infant mortality rates were among the highest in the Western world. At that same time, a significant push was under way throughout the country to replace lay midwives with trained physicians. Mary Carson Breckinridge (1881–1965), a member of a politically prominent Kentucky family, shared the widespread concern that American women, particularly women living in poor, rural areas, lacked access to medical care. After the deaths of her own two small children, she pledged to assist other women and their babies through the creation of a public health organization. In 1925 she established the Frontier Nursing Service (FNS) in Leslie County, Ky., to demonstrate how medical care could be made accessible and affordable. Initially called the Kentucky Committee for Mothers and Babies, Breckinridge's organization pioneered a new model of health care, based largely on European ideas.

Breckinridge's innovative plan placed nurse-midwives in a central role. Because no American schools offered specialized training in midwifery until

Mary Breckinridge at her home, the Big House, with the 10,000th Frontier Nursing Service baby, Marlene Wooton, during the first Mary Breckinridge Day, 1962 (Courtesy Frontier Nursing Service, Wendover, Ky.)

the 1940s, Breckinridge recruited the organization's first employees from Great Britain, where she had also earned her midwifery certificate. Between 1925 and 1930, the FNS built a 12-bed hospital in Hyden, Ky., and eight outpost centers to serve the needs of approximately 10,000 people residing in the service's territory, which eventually included parts of Clay, Perry, and Harlan counties. Although Breckinridge's stated goal was simply to deliver healthy babies, the organization's mission expanded quickly to emphasize preventative care, including dental and eye checkups, for all members of the families it served. Breckinridge also looked for ways to strengthen the region's economy, recognizing the significant link between health and financial hardship.

Breckinridge used her extensive family contacts and name recognition to build a national following for her work, and a network of donors generously funded her organization. The FNS received very little public funding during its founder's lifetime. Romantic tales of the FNS's work appeared frequently in mass-circulation magazines beginning in the late 1920s. Journalists praised FNS nurses' willingness to travel over rough terrain any time of the day or night to bring a new life into the world. In the public mind, Frontier Nurses became famed "angels on horseback," risking life and limb to serve their patients.

From the start, Breckinridge hoped that her demonstration would be replicated elsewhere, and the FNS itself attempted to spin off satellite branches in other areas of the country, but the challenges of the Great Depression interfered with this plan. World War II added difficulties as the bulk of the service's foreign-born staff returned home. In 1939 the FNS created a midwifery school to train new nurse-midwives, and it is through this avenue that Breckinridge's ideas have had their greatest reach both within the United States and worldwide. At the time of Breckinridge's death in 1965, FNS nurses had delivered over 14,500 babies, treated nearly 58,000 patients, and given 248,000 inoculations. Breckinridge left the organization with an endowment topping $2 million.

After the founder's death, Helen Browne was named director. She led the organization through a period of expansion, adapting the service to accept Medicare and Medicaid funding and raising nearly $7 million to fund construction of the 40-bed Mary Breckinridge Hospital. Named a Critical Access facility in 2003, the hospital serves a four-county area. As well, the organization, which continues to rely heavily on private contributions, runs a home health agency and four rural health-care clinics. The Frontier School of Midwifery and Family Nursing continues to produce nurse-midwives, family nurse practitioners, and women's health-care nurse practitioners through distance education programs that allow students to remain in their home communities while they train. By 2010 the school reported more than 2,000 graduates, representing all 50 states.

MELANIE BEALS GOAN
Lexington, Kentucky

Mary Breckinridge, *Wide Neighborhoods: A Story of the Frontier Nursing Service*

(1981); Frontier Nursing Service, Inc., www.frontiernursing.org (2011); Frontier School of Midwifery and Family Nursing, www.frontierschool.edu (2011); Melanie Beals Goan, *Mary Breckinridge: The Frontier Nursing Service and Rural Health in Appalachia* (2008).

Garden, Alexander

(1730–1791) PHYSICIAN AND NATURALIST.

Alexander Garden was born at Birse, near Aberdeen, where his father was a Church of Scotland minister. He attended Marischal College and began the study of medicine as an apprentice to an Aberdeen physician. He served two years in the Royal Navy as a surgeon's mate. After study at the University of Edinburgh, in 1752 he immigrated to South Carolina, becoming one of the leading physicians of Charleston. Although Garden was highly regarded there, his loyalty to Great Britain necessitated that he leave when Loyalists were expelled after the Revolution. He spent the remainder of his life in London.

Although he has sometimes been called the most famous physician of colonial times, Garden made even greater contributions as an untiring student of natural history. Very shortly after his arrival in South Carolina he established himself as one of the small group of men who were eagerly helping European scientists to understand the natural history of North America. He was ambitious for recognition as a scientist of importance. His interests were broad, including plants, animals, medical problems, the health of slaves, climate, minerals, and other aspects of the environment. He sent plant specimens to Charles Alston in Edinburgh, John Ellis in London, and Carolus Linnaeus in Sweden. Amphibians, fish, insects, and reptiles went to Linnaeus and to John F. Gronovius in Holland. Birds were supplied to Thomas Pennant in England. Garden contributed to the local scene as well, being a stimulus to, and an active participant in, the intellectual and cultural life of Charleston.

His importance was well recognized in his own time. He was elected to membership in the most prestigious scientific societies in both America and Europe, including the American Philosophical Society, the Royal Society of Arts and Science of Uppsala, the Royal Society of London, the Royal Society of Arts (London), and the Royal Society of Edinburgh. Linnaeus named the gardenia flower in his honor.

Garden's publications were few, largely journal articles describing plants and animals. Many of his letters survive and testify to his facility as a writer. His correspondents included all of those persons mentioned above and also John Bartram, Cadwallader Colden, John Clayton, and Benjamin Franklin.

EDMUND BERKELEY
DOROTHY SMITH BERKELEY
Charlottesville, Virginia

Edmund Berkeley and Dorothy Smith Berkeley, *Dr. Alexander Garden of Charles Town* (1969); David Taylor and Rudy Mancke, eds., *South Carolina Naturalists: An Anthology, 1700–1860* (1998).

Geophagia and Pica

Geophagia (or geophagy) is the intentional consumption of earth. In Western culture, it is commonly labeled a pica, an eating disorder characterized by the consistent consumption of nonnutritive substances such as feces (coprophagy), starch (amylophagy), or wood (xylophagia). In 2000 a panel convened by the Agency for Toxic Substances and Disease Registry in Atlanta proposed to distinguish between geophagia (culturally sanctioned behavior) and soil pica ("recurrent ingestion of unusually high amounts of soil . . . on the order of 1,000–5,000 milligrams per day"). Other differences between both behaviors include impacted populations (pica is more prevalent among children under six, geophagia is reportedly more prevalent among pregnant women), consumed material (surface soil, possibly contaminated, for pica versus clays found at a 18- to 36-inch depth), and health consequences (generally detrimental for pica and potentially beneficial for geophagia).

A universal though limited practice, geophagia has been documented in the United States, especially among African Americans in the South and American Indians in the Southwest. Its introduction is commonly attributed to African slaves, an erroneous assumption since American Indians practiced geophagia before the arrival of European settlers.

Although individuals of both sexes of all races and ages practice geophagia and pica, they are more frequent among particular categories, especially young children, southern women, pregnant women, and persons with mental disabilities. For instance, documented geophagia in several southern states (Alabama, Georgia, Louisiana, Mississippi, North Carolina, and Tennessee) occurs almost exclusively among rural African American females. They reportedly have been introduced to the practice by relatives and have, in turn, socialized their children to it.

Its assumed origin and practitioners have combined to ascribe geophagia a minority status in the United States and Western culture in general. Rurality, being female, and being black are traits traditionally associated with poverty, especially during the period (from the 1930s to the 1970s) when most studies of geophagia in the United States were conducted. In addition, a correlation between pica and malnutrition and reported low levels of education among dirt eaters contributed to tie the practice to poverty and low social class.

So why do people eat dirt? Explanations of pica and geophagia remain tentative at best, even though along the years psychological, cultural, nutritional, physiological, and sensory hypotheses have been advanced. While pica is universally and consistently labeled a pathology—a mental or eating disorder—geophagia is treated as both adaptive strategy and pathology. Iron- and calcium-deficiency have been offered as factors of geophagia among pregnant women, who would find in clay much needed minerals. Other rationales include calming hunger pains, alleviating malnutrition, or being socialized to the practice.

Although the incidence of geophagia is difficult to measure, it is assumed to be declining. The urbanization of American society has greatly limited access to sources of dirt. Greater availability of health care and commercialization of nutritional supplements have arguably limited the need or compulsion to resort to eating dirt as a health regimen. Finally, the stigmatization of the practice, especially by the medical profession, has certainly limited its transmission.

JACQUES HENRY
University of Louisiana at Lafayette

Agency for Toxic Substances and Disease Registry, *Summary Report for the ATSDR Soil-Pica Workshop* (2001); Gerald N. Callaghan, *Emerging Infectious Diseases* (August 2003); P. Wenzel Geissler, *Africa* (2000); Jacques Henry and Alicia Matthews Kwong, *Deviant Behavior* (July–August 2005); Ran Knishinsky, *The Clay Cure: Natural Healing from the Earth* (1998); Ella P. Lacey, *Public Health Reports* (January–February 1990); Russell M. Reid, *Medical Anthropology* (1992); Donald Vermeer and Dennis A. Frate, *American Journal of Clinical Nutrition* (1979).

Guyton, Arthur C.

(1919–2003) EDUCATOR AND SCHOLAR.

Dr. Arthur C. Guyton spent his career as chairman of the Department of Physiology and Biophysics at the University of Mississippi Medical Center in Jackson. During those five decades, he became one of the most highly regarded physiologists in the world and the author of the *Textbook of Medical Physiology*, in publication since 1956, a medical textbook of unparalleled popularity. He also wrote 40 other books and more than 600 articles for scientific publications. His work led to a new understanding of the cardiovascular system and changed the entire field of physiology.

Guyton was born in Oxford, Miss., in 1919, the son of Dr. Billy S. Guyton and Mary Katherine Smallwood. He graduated from the University of Mississippi and went on to Harvard Medical School, where he earned the M.D. degree in 1943. After medical school, he married Ruth Weigle, the daughter of the dean of the Yale Divinity School and chairman of the committee responsible for the Revised Standard Version of the Bible. They had 10 children, all of whom became physicians.

After a year's internship at Massachusetts General Hospital in Boston, he served in the U.S. Navy for two years at the National Naval Medical Center in Bethesda, Md., and at Camp Detrick in Maryland. Back in Boston to complete a residency in surgery at Massachusetts General, Guyton contracted polio in 1946. He and Ruth spent months at Warm Springs, Ga., where he regained the use of some of his paralyzed muscles but not all. He had residual paralysis in the right lower leg, left upper arm, and both shoulders.

"It was clear," Guyton wrote, "that I couldn't be a surgeon as I had planned. But that meant that I could devote myself to the two things that meant the most to me: medical research and raising a family." The Guytons, with the first 2 of their 10 children, moved back to Oxford in 1947 where he began teaching physiology and doing research. He was named chairman of physiology in 1948 and moved to Jackson in

1955 when the University of Mississippi Medical Center opened and the school expanded to a full four-year curriculum.

In 1956 the first edition of his textbook was published, and he received a Presidential Citation for devices he designed for those disabled by polio—an electronic wheelchair (which he never used), a special hoist for moving patients from bed to chair, and an automatic locking and unlocking leg brace. At the same time, he was hard at work on his quest to discover the cause of hypertension, or high blood pressure. In his search, he discovered that most of what had been written about the heart, blood flow, blood vessels, and blood pressure control was wrong.

His first major breakthrough was the discovery that cardiac output—the amount of blood pumped by the heart—depended not on the heart but on the demand of tissues for oxygen. This theory significantly advanced the understanding of blood circulation, fully explained in his book *Circulatory Physiology: Cardiac Output and Its Regulation*, published by W. B. Saunders in 1963.

He filled in another missing link in the understanding of cardiovascular physiology when he proposed, to great opposition in the physiology community, that the pressure in the interstitial fluid was negative. The interstitium is the space between cells, and understanding its characteristics was vital to understanding edema and congestive heart failure. Guyton proved his theory by devising perforated plastic capsules, surgically implanted in the interstitium.

With some of the basic mechanisms of the circulation worked out, he turned to blood pressure again. Working with an analog computer and the first computer model of the circulatory system, he found that the only factor that could control blood pressure long term was fluid control by the kidney. It flew in the face of prevailing notions of blood pressure control, but was eventually accepted by the scientific community. It has been the basis of much drug development in the treatment of hypertension.

Guyton received nearly every conceivable prize the field of physiology can bestow. The one prize, however, that clearly demonstrates the importance of his contributions was the invitation by the Royal College of Physicians in London to give the William Harvey Lecture at the 400th Anniversary Symposium in 1978 commemorating the birth of Harvey, the scientist who first described the circulation of blood.

With his research and teaching, Guyton changed physiology from a science of verbal descriptions to one of quantitative analysis. He brought mathematics and physics into the discipline. He was a pioneer in the use of computers to study body function and has taught scientists all over the world computer simulation. A descendant of his original computer model of the cardiovascular system is now used by NASA scientists to determine ways to counter the effects of weightlessness on astronauts.

It is difficult to overestimate the influence of his famous *Textbook of*

Medical Physiology, which has been through 11 editions and translated into 12 languages. The Association of American Medical Colleges awarded Guyton the Abraham Flexner Award for Distinguished Service to Medical Education in 1996 for the textbook, which is used in more medical schools around the world than any other physiology text.

Guyton and his wife died in 2003, the result of injuries sustained in an automobile accident. Guyton died at the scene, and Ruth died a week later.

JANIS QUINN
University of Mississippi Medical Center

Carol Brinson and Janis Quinn, *Arthur C. Guyton: His Life, His Family, His Achievements* (1989); Arthur C. Guyton, "A Brief History of Cardiovascular Physiology at Mississippi," unpublished manuscript in the Department of Physiology and Biophysics, University of Mississippi Medical Center, videotaped interview for the Rowland Medical Library Oral History Project, RML archives, University of Mississippi Medical Center (16 November 2001); Janis Quinn, *This Week at UMC* (30 November 1996, 22 September 2000).

Hardy, James D.

(1918–2003) SURGEON.

Dr. James D. Hardy was the first chairman of the Department of Surgery at the University of Mississippi Medical Center from the opening of the Medical Center in Jackson in 1955 until his retirement in 1987. He and a team of surgeons from the Medical Center performed the world's first lung transplant in 1963; the patient lived for 18 days before dying of kidney failure. Then, in 1964, Hardy and his team took the heart from a chimpanzee and transplanted it to a man dying of heart disease, the world's first heart transplant. Hardy is generally credited with having performed the first lung transplant, but rarely for having performed the first heart transplant, which predated the human-to-human heart transplant by Dr. Christiaan Barnard in South Africa by three years. Barnard's patient lived 18 days.

Hardy grew up in Newala, Ala., about 35 miles south of Birmingham, where his father owned a lime manufacturing plant. He graduated from the University of Alabama and entered medical school at the University of Pennsylvania. He earned his medical degree, served in the military during World War II, and entered surgical residency at Penn, finishing in 1951. He was serving as the director of surgical research at the University of Tennessee when Dr. David Pankratz, dean of the School of Medicine at the University of Mississippi, tapped him in 1953 to be the chairman of surgery at the University of Mississippi Medical Center, an institution still on the drawing board. Hardy, by then married to Louise Scott Sams and the father of four daughters, moved his family to Jackson in 1955 in time to oversee the department's formation in the new Medical Center.

Recognizing that organ transplantation was the next big advancement in surgery, Hardy began equipping labs for transplant research. From 1956 to 1963, Hardy, along with Drs. Watts Webb, Martin Dalton, and Fikri Alican, performed nearly 1,000 lung transplants in animals. His decision to move to human transplantation was made only after the drug Imuran had been reported to impede the body's rejection of the transplanted organ. Hardy began using the drug in his animal labs and found that a transplanted lung could last at least 30 days if he used Imuran. With the clear possibility of a transplanted lung lasting that long, Hardy now had justification for its clinical use. He had seen many patients while unconscious vomit the contents of their stomachs and suck it into their lungs. The stomach acids paralyzed the lungs temporarily. The lungs could heal with time, but the patients often died before the lungs had a chance to heal. A transplanted lung, even if it lasted only a few weeks, would allow time for healing. And even if it was eventually rejected, the patient would be left with one healthy lung.

He sought permission from the Medical Center's vice chancellor, Dr. Robert Marston, to proceed to human lung transplantation, assuring Marston that the team would follow prescribed guidelines. First, the recipient would almost certainly have a fatal disease. If the operation went sour, it could not be argued that the transplant shortened his or her life. Second, there must be a reasonable possibility that the patient would benefit from the transplant. Removal of the patient's diseased lung would not result in the sacrifice of a significant amount of the patient's own functioning lung tissue. Third, the lung to be replaced would be the left lung because it has longer, unbranched portions of the pulmonary artery and main bronchus and was technically easier than the right lung.

John Russell, a patient who met all the guidelines, was admitted to University Hospital on 15 April 1963. He agreed to the surgery, and two months later a donor lung became available. Even though Russell lived only 18 days after the transplant, Hardy considered the operation a success. The lung was still functioning with no signs of rejection at his death.

The transplant attracted little local notice because of another, more tragic event that occurred the same night. As Hardy was closing on Russell, the team received an urgent call to report to the emergency room. Hardy asked Dalton to go to the ER while he closed. Dalton rushed downstairs to find a young black man who had been shot at close range by a rifle. Dalton attempted to stop the bleeding, using manual heart massage and defibrillation—to no avail. Thirty minutes later Dalton pronounced death. Only after he told those gathered in the waiting room the sad news did he learn that the dead man was Medgar Evers, the civil rights activist.

With the success of the lung, Hardy wanted now to do a human heart transplant. He again got permission from Marston and readied two surgical teams to take care of both the donor and the recipient. On 23 January 1964 Boyd Rush, the patient Hardy had identified as a heart recipient, was dying of heart failure. A patient in the hospital's intensive care unit, breathing only with the mechanical support, was identified as a possible donor. But in the days before brain death classification, Hardy's team had to wait until the heart stopped beating. Hardy was trying to keep Rush alive, but the donor's heart was still beating vigorously. Hardy made the decision to use a chimpanzee's heart instead. It beat for 90 minutes before failing.

Hardy had prepared himself for a certain amount of criticism, but he thought most of it would come from the general public. He did not count on the outcry from his colleagues. He was maligned at national surgical meetings, his clinical integrity questioned. He called it a "searing experience." The criticism abated when his paper in the *Journal of the American Medical Association* appeared in June 1964 wherein he detailed the strict guidelines he used in selecting both donor and recipient, his work in the labs leading up to transplant, and the strong scientific basis for the act.

Hardy also amassed a scholarly record that is rarely equaled. He wrote or edited 23 books, including two autobiographies, and he published nearly 600 papers in medical journals. He served as president of every major surgical society in the world.

A national moratorium on organ transplantation followed Hardy's heart transplant because doctors generally recognized that organ rejection would have to be overcome if transplantation were to become the viable clinical option it has become. Hardy's giant steps in the early era of transplantation helped create that option.

JANIS QUINN
University of Mississippi Medical Center

Martin Dalton, *Annals of Thoracic Surgery* (November 1995); Mary Jo Festle, *Journal of Mississippi History* (Summer 2002); James D. Hardy, *The World of Surgery, 1945–1985: Memoirs of One Participant* (1986), in *Jonathan E. Rhoads' Eightieth Birthday Symposium* (1989); Jürgen Thorwald, *The Patients* (1971).

Herty, Charles Holmes

(1867–1938) CHEMIST.

Charles Holmes Herty was a southern chemist whose career as a teacher, researcher, and publicist contributed significantly to the economic development of the South and the growth of chemistry and the organic chemical industry in the United States. Born in Milledgeville, Ga., and educated at Johns Hopkins (Ph.D. 1890), Herty taught at the universities of Georgia and North Carolina from 1891 to 1901 and 1905 to 1916, respectively. Associated with the Bureau of Forestry in the U.S. Department of Agriculture between 1902 and 1904, he patented the Herty Cup and Gutter System of turpentining, which, by supplanting the destructive system of "boxing," then universally practiced in the American South, helped to save the threatened naval stores industry.

Twice elected president of the American Chemical Society (1915–16), Herty attained a national reputation as an expert in wood chemistry; and his articulate advocacy of corporate, academic, and governmental cooperation in the establishment of chemical independence from Europe made him a natural for the editorship of the American Chemical Society's *Journal of Industrial and Engineering Chem-*

istry. He served in that capacity for four years (1917–21), with time out for trips to Europe to advise the American peace commissioners at Versailles on matters regarding German dyestuffs and German patents seized during the war. In 1921 he became president of the Synthetic Organic Chemical Manufacturers' Association, continuing his close relationship with Francis P. Garvan, alien property custodian in the Wilson administration and president of the Chemical Foundation, created in 1919 to administer the development by private American firms of the seized German patents. In 1926 Herty left the manufacturers' association, became a consultant to the Chemical Foundation, and for two years worked closely with Garvan and foundation funds to promote research in chemotherapy. He also ran his own consulting firm in New York.

Finally, still interested in southern pine trees, conservation, and diversification of the South's stagnant economy, Herty attracted enough funds from the Chemical Foundation, the state of Georgia, and the citizenry of Savannah to establish an experimental pulp and paper laboratory there in 1932. His goal was to prove that sulfite pulp, suitable for the manufacture of fine white newsprint, could be economically made from young southern pines, thereby encouraging reforestation and freeing the U.S. newspaper industry of its dependence on the more expensive and slow-growing spruce pulp imported largely from Canada. At the same time, the poor southern farmer, plagued with worn-out fields or no market for his cotton, would be provided with an alternative cash crop. Most important, new industry would be attracted to the region. Herty's papermaking demonstration, conducted under commercial conditions in a Canadian newsprint factory with pulp shipped from the Savannah lab, was a thorough success. Nine southern dailies printed their 20 November 1933 editions on the first run. As a direct result of Herty's efforts, although he did not live to see it, the first southern newspaper plant was erected near Lufkin, Tex. By 1955 an expanded plant at Lufkin, together with two more firms established in Alabama and Tennessee, accounted for 29 percent of all newsprint produced in the United States.

GERMAINE M. REED
Georgia Institute of Technology

Charles Holmes Herty Papers, Special Collections, Robert Woodruff Library, Emory University; Jack P. Oden, *Journal of Forest History* (April 1977); Gerry Reed, *Crusading for Chemistry: The Professional Career of Charles Holmes Herty* (1995), *Georgia Historical Quarterly* (Fall 1993), *Journal of Forest History* (October 1982).

HIV/AIDS

Acquired immunodeficiency syndrome (AIDS) is the final stage of a communicable disease caused by the human immunodeficiency virus (HIV). HIV/AIDS cripples the immune system and prevents its host from combating other diseases. The virus spreads via sexual intercourse, intravenous needles, perinatal transmission, infected breast milk, and contaminated blood and tissue. The disease arrived in the United States in the late 1960s but was not discovered until

1981. By the end of 2010, HIV/AIDS had killed more than 600,000 Americans, including 230,000 southerners. An additional 1.2 million Americans were living with the infection at that time.

By the end of the epidemic's first year, almost every state in the South had reported indigenous cases. As elsewhere in the country, homosexuals in southern cities were the first group noticeably affected by the disease. In response, southern gays organized care, prevention, and education efforts in their communities; as early as 1982, AIDS organizations emerged in Texas, Georgia, and North Carolina. Many leaders from these groups went on to direct regional or statewide responses to the epidemic. In communities lacking a public gay community, concerned health professionals, community groups, and religious organizations spearheaded AIDS efforts.

Southern research institutions played a prominent role fighting AIDS. In 1984, one year after French researchers identified the suspected microorganism, scientists at Maryland's National Cancer Institute (NCI) demonstrated that it caused AIDS and produced a blood test to help diagnose infection. In 1985 researchers at Burroughs Wellcome in North Carolina found that the drug azidothymidine (AZT) inhibited viral replication. Consequently, the Food and Drug Administration (FDA) rushed the drug through approval and onto the market by March 1987. The drug was quite toxic, however, and Burroughs Wellcome's high initial price sparked congressional hearings, prompted federal drug assistance programs, and drove outraged activists to protest high drug prices and the lethargic pace of drug development through the AIDS Coalition to Unleash Power (ACT UP).

By the early 1990s, it also became clear that AZT's effects were short-lived. Consequently, researchers continued searching for new and combined therapies. In 1996 a new therapy that combined older drugs with protease inhibitors—called highly active antiretroviral therapy (HAART)—rendered HIV/AIDS a manageable disease by allowing clinicians to reduce viral counts to negligible levels and markedly cut AIDS-related mortality. Concurrently, researchers in North Carolina, Florida, and elsewhere found they could significantly reduce mother-to-child-transmission by treating pregnant women with AZT.

The socially taboo ways HIV spread—sexual transmission, multiple partners, male homosexuality, injection drug use—made AIDS especially contentious in the socially conservative South. Almost every southern state had antisodomy laws in place before 1992; more than half retained them until 2003. Many religious conservatives, like Virginia's Jerry Falwell, associated AIDS with divine retribution. Many political conservatives, like U.S. senator Jesse Helms in North Carolina, opposed expending public funds on conducting prevention among marginalized groups. Many communities were divided over condom promotion in schools, explicit content in prevention material, and harm-reduction efforts among drug addicts. HIV-related fears proliferated: infected workers lost jobs, infected children were barred from some schools, gay men lost insurance or parental rights, and infected families were ostracized and their houses vandalized. Still, many communities and organizations, like the Lambda Defense and Education Fund, mobilized to defend the rights of those with HIV. By the end of the 1980s, a few southern states had even instituted antidiscrimination laws for people with HIV, a fact that became institutionalized nationally through the 1990 federal Americans with Disabilities Act (ADA).

AIDS in the South had a particular shape relative to the rest of the nation. AIDS disproportionately affected blacks from the very first cases. By the 2000s blacks constituted more than 50 percent of new infections, though they composed only 18 percent of the population. These disparities stemmed from several factors, including high rates of sexually transmitted diseases, lower overall health access and outcomes, increased socioeconomic vulnerability, and segregated sexual networks. Rural communities were also extremely vulnerable, as rural physicians, including Abraham Verghese in Tennessee and Richard Rumley in North Carolina, documented. Thus, by the early 2000s, southern states accounted for 7 of the 10 states with the highest AIDS case rates, and the South reported a disproportionate HIV prevalence among young adults. Despite this skewed impact, federal funding for AIDS (the Ryan White CARE Act) still favored other regions of the country throughout the 1990s and early 2000s.

In light of these trends, AIDS directors from southern states joined together in 2001 to address the growing disparity of AIDS in the South. Dubbed the Southern AIDS Coalition (SAC), the group in 2003 published its "Southern States Manifesto" to highlight the disproportionate number of new HIV infections in the South, the disease's impact on minorities, and the inequitable distribution of federal funds in light of the region's AIDS burden. SAC's advocacy efforts to obtain increased federal funding met with success in 2006 with the passage of the Ryan White HIV/AIDS Treatment Modernization Act, which made the original legislation more responsive to southern needs. However, the recession and government austerity measures at the end of the decade created renewed concern that the infrastructural advances many southern states were making would be imperiled. Indeed, the story of AIDS in the American South symbolizes the much larger problem of health disparities in the South, as the region's poor health infrastructure, entrenched poverty, poor overall health outcomes, and racial disparities serve to leave many of its population vulnerable to disease and untimely death.

STEPHEN INRIG
Southwestern Medical Center
University of Texas

Stephen Inrig, *North Carolina and the Problem of AIDS: Advocacy, Politics, and Race in the South* (2011); Jacob Levenson, *The Secret Epidemic: The Story of AIDS and Black America* (2004); Abraham Verghese, *My Own Country: A Doctor's Story* (1994); Kathryn Whetten-Goldstein and Trang Quyen Nguyen, *"You're the First One I've Told": New Faces of HIV in the South* (2003).

Hookworm

The parasitic nematode worm *Necator americanus*, or the New World hookworm, was found to infect a significant portion of the population in the southern United States during the first decade of the 20th century. Hookworms thrive in warmer climates and sandy soils, entering hosts through the skin and eventually ensconcing themselves in the intestinal wall and living off the victim's blood supply. There the nematode typically enjoys a five-year life-span, laying up to 10,000 eggs daily, which exit the body in fecal matter. Victims often harbor multiple hookworms, a condition commonly leading to uncinariasis, or hookworm disease, and causing anemia, fatigue, weakness, and strange cravings (including those associated with pica, or dirt eating) and potentially stunting mental and physical growth in children. Ignorance of the disease and its mechanisms, coupled with an environment hospitable to the worm, the prevalence of bare feet, "open-air" defecation, and a paucity of sanitary waste disposal systems led to its endemic presence in the South at the turn of the century.

The condition was deemed the "germ of laziness" not long after its discovery by Charles Stiles in 1902 and linked to widespread listlessness among southerners. In 1905 reports indicated up to 40 percent of school-aged children in the region harbored the parasite. These startling results prompted John D. Rockefeller to develop and fund the Rockefeller Sanitary Commission for the Eradication of Hookworm Disease (RSC). Between 1910 and 1915, the commission undertook a campaign to diagnose, treat, and educate the public in 11 southern states, dispensing Thymol, a thyme-based dewormer, free of charge and helping communities to construct sanitary latrines. Though the RSC did not fully eradicate hookworm, its efforts contributed to a significant decline in regional infection rates. Likewise, the commission spearheaded the development of county health boards, creating a much-needed institutional presence capable of continuing the hookworm campaign after the withdrawal of the RSC, as well as monitoring sundry communal health issues and addressing other regional scourges, including malaria and pellagra.

Unfortunately, public awareness of the pervasiveness of hookworm in the South helped to reinforce negative stereotypes concerning the hygiene and work ethic of its denizens in the early 1900s. However, champions of the RSC project also challenged arguments touting the beneficial aspects of eugenics initiatives targeting the South, claiming its citizenry suffered from chronic and ubiquitous disease rather than genetic inferiority. Additionally, proponents dovetailed their interests with those of the New South, suggesting worker productivity would improve with the eradication of hookworm, to the benefit of the region's burgeoning industrial sector. While the long-term effects of the RSC project on southern industry remain unavailable at present, Hoyt Bleakley's 2007 study indicates higher school attendance, literacy rates, and eventual income levels among children treated during the RSC campaign in areas severely afflicted by the parasite.

KATHRYN RADISHOFSKI
University of Mississippi

Hoyt Bleakley, *Disease and Development: Evidence from Hookworm Eradication in the American South* (2007); John Ettling, *The Germ of Laziness: Rockefeller Philanthropy and Public Health in the New South* (1981); Matt Wray, *Not Quite White: White Trash and the Boundaries of Whiteness* (2006).

Hoxsey Therapy

A naturopath and purveyor of one of the most popular alternative medical therapies in United States during the 20th century, Harry Hoxsey administered treatment to thousands of cancer patients at his Dallas, Tex., clinic between 1936 and 1958. Hoxsey waged a populist war against the strictures and stronghold of the official U.S. medical establishment and institutions like the American Medical Association (AMA), believing his holistic, herb-based cancer treatment to be superior to the more invasive cancer remedies that were deployed by standard practitioners, such as chemotherapy, radiation, and surgery. Indeed, Hoxsey claimed his therapy cured 85 percent of patients afflicted with external cancers, and 25 percent of those suffering internal malignancies, declaring in 1954, "I can cure cancer. I have saved thousands of lives; can save millions more. And I challenge organized medicine to let me prove it." However, because Hoxsey lacked an accredited medical license and proof of the medical value of his formulas beyond patient testimony, the FDA deemed Hoxsey a quack, targeting his clinics with legislation and judicial orders that eventually forced him out of business.

Hoxsey's regimen included a purgative tonic containing herbal ingredients such as licorice, red clover, burdock root, and prickly ash; dietary restrictions; and the application of topical salves. The overall efficacy of Hoxsey's cancer therapy proved difficult to determine by standard evaluation on account of his clinic's unorthodox and incomplete methods of documentation. According to the AMA, though Hoxsey's arsenic and bloodroot-based salves worked as corrosive escharotics, known and utilized with some success against external cancers since the 19th century, surgery remains the recommended treatment in such cases and the preferable option for patients, as escharotics can be both painful and disfiguring. Interestingly, the American Cancer Society contends that, although clinical trials on mice did not show Hoxsey's herbal tonic to have any positive effect on cancerous tumors, recent studies indicate some of the individual ingredients may be useful in treating the disease.

Harry Hoxsey, an Illinois native, was born 23 October 1901, the youngest of 12 children. He purportedly obtained his remedial formulas from his father, John Hoxsey, a veterinarian who earned extra income as a lay practitioner attending to local cancer patients. John's Kentuckian grandfather allegedly developed the therapeutic recipes after witnessing a horse with a cancerous lesion fully recover from the illness while subsisting on an herb-laden diet. Hoxsey maintained his own career in medicine began at the behest of a Civil War veteran suffering from cancer, who pleaded with Harry to treat him in spite of Hoxsey's lack of credentials. Harry later asserted the AMA interfered with his acceptance to medical school because of his history of administering treatment without a license.

Harry finally settled in Dallas in the mid-1930s after years of abortive attempts to establish clinics throughout the Midwest. There, Hoxsey earned his naturopathic license and recruited a host of influential supporters, including the local district attorney, a Texas federal judge (who later ruled in Hoxsey's favor in a libel suit against Hearst newspapers and Morris Fishbein, editor of *JAMA: The Journal of the American Medical Association* and Hoxsey's most outspoken and high-profile critic), several U.S. senators, a cadre of medical outsiders invited to conduct an investigation of the Hoxsey clinic, the popular evangelist Gerald Winrod, and the American Rally, a conservative political party. Though still not invulnerable to the arrest warrants and injunctions ultimately foreclosing the viability of his earlier efforts outside the South, Hoxsey transformed his Dallas clinic into a lucrative enterprise, eventually netting more than a $100,000 in annual revenue and encouraging Hoxsey to the open additional clinics in Pennsylvania, Denver, and Los Angeles. Notably, Hoxsey was among the few practitioners in Dallas willing to treat African Americans at the time and likewise took on many low-income patients free of charge.

Though the early efforts of the FDA and Morris Fishbein seemed to have little effect on the success of Hoxsey's Dallas operation, their scrutiny, along with that of the National Cancer Institute, eventually resulted in the production of public notices warning cancer patients against his treatments. By the late 1950s, federal and state injunctions sealed Hoxsey's fate, and his Dallas clinic closed in 1958. Five years later, Hoxsey's chosen successor and former nurse, Mildred Nelson, opened the Bio-Medical center in Tijuana, Mexico, beyond the jurisdiction of the U.S. medical establishment; as of 2011, the clinic still administers the Hoxsey therapy, though Nelson died in 1999. Ironically, Hoxsey developed prostate cancer in the last half of the 1960s, seeking surgery and conventional treatment after his own tonic failed to produce results. Hoxsey died in 1974 in relative obscurity.

KATHRYN RADISHOFSKI
University of Mississippi

Kenny Ausubel, *When Healing Becomes a Crime: The Amazing Story of the Hoxsey Cancer Clinics and the Return of Alternative Therapies* (2000); Eric Juhnke, *Quacks and Crusaders: The Fabulous Careers of John Brinkley, Norman Baker, and Harry Hoxsey* (2002).

Influenza Epidemic of 1918

Known variously as "Spanish influenza," "the Spanish lady," or "purple death" (for the characteristic coloring of the oxygen-starved victims), the influenza epidemic of 1918 claimed more American lives than World War I. Victims often progressed rapidly from mild illness to death within a few days. Unlike most flu strains, which tended to result in death primarily among the elderly and those already in poor health, this epidemic predominantly targeted infants and otherwise healthy adults between 25 and 35.

Southerners seemed particularly susceptible, for several reasons. First, the health-care system in the South was less developed than in other areas of the country, and a significant number of southern medical professionals were called to duty in the war. Second, the majority of southerners lived in rural areas, where they were less likely to have been exposed to strains of the flu that would have allowed them to develop some level of immunity. Third, the war sent many southerners to military camps and industrial centers, where overcrowded conditions contributed to the rapid spread of disease. And fourth, those who became ill and returned home to their rural communities spread the disease into areas least equipped to handle it.

Many date the first outbreak of the influenza epidemic to March 1918, with the case of an army cook at Camp Funston, in Fort Riley, Kans. However, there are also accounts of an earlier outbreak in a nearby Haskell County, Kans., community, as well as an outbreak of "Texas colds" at Fort Sam Houston in February. The main part of the epidemic hit most areas of the South in September 1918. By the end of the year, it had killed more than 6,000 people in Mississippi and more than 9,000 in neighboring Alabama. More than 3,000 deaths from influenza were recorded in Louisiana in just the short span of time between 28 September and 5 November; more than half of these deaths were in New Orleans. The rest of the South experienced similar losses. Undoubtedly, the epidemic was underreported, particularly since the overburdened medical professionals were more focused on getting through the epidemic rather than reporting it.

The rapid progress of the epidemic made caring for the sick and preventing the spread of the disease extremely difficult. In some areas, hundreds of new cases of influenza were reported each day. There were not enough medical personnel and hospital beds to meet the demands, and racial segregation further complicated matters by necessitating separate facilities for African American and Latino patients.

People were not sure how to recognize the illness or how to prevent it, and they were panicked by the rapid onset of the disease, the tendency for the disease to progress to pneumonia, and the increasing number of deaths. Whole families became ill at once, and neighbors rendering aid were likely to become ill themselves and further spread the disease. Parents began to demand the return of their children from schools, either out of fear for their health or out of a need for help in nursing family members at home who had become ill.

Luckily, though the war had probably contributed to the spread of the disease, it also contributed to a proliferation of organizations poised to help in a crisis. Schools were pressed into service as makeshift hospitals. Women's clubs and community organizations (such as the Nashville Golf and Country Club) helped with childcare, transportation, and meal preparation. Large employers, such as Pacific Mills in South Carolina, organized food deliveries for the families of their workers.

In most places, health officers closed public meetings, public schools, and entertainment venues in towns where cases of influenza appeared, and state and county fairs were suspended. In Mississippi, the State Board of Health restricted public gatherings further by ordering private funerals and banning the practice of taking bodies into churches. Many people wore masks in public. In some places, warning posters or bright yellow flags were placed on infected homes. The town of Quitman, Ga., instituted no fewer than 27 regulations designed to combat the spread of influenza.

The lack of a clear solution to the problem resulted in a rash of unproved "remedies." Store supplies of Vicks Vaporub were wiped out, and one traditional concoction of whiskey and lemons drove the price of lemons up from $3.50 to $13.50 per box in Louisiana and helped to cause a nationwide shortage of the fruit. The situation was serious enough that some doctors began prescribing alcohol as a treatment for pneumonia, causing state officials to release confiscated beverages for medicinal use in the face of impending national prohibition.

The disease had begun to abate in many communities by the end of October, though it lingered on or recurred in others during the early months of 1919. Clergy, school boards, and business owners pushed for the repeal of health regulations restricting social interactions. With both the war and the influenza epidemic over, southerners got back to their normal lives, though none who lived through it would ever forget that time.

DIANE DECESARE ROSS
University of Southern Mississippi

John M. Barry, *The Great Influenza: The Epic Story of the Deadliest Plague in History* (2004); Allan D. Charles, *Journal of the South Carolina Medical Association* (August 1977); Alfred W. Crosby, *America's Forgotten Pandemic: The Influenza of 1918* (2003); Patricia D'Antonio and Sandra B. Lewenson, *History as Evidence: Nursing Interventions through Time* (2011); Geraldine M. Emerson, *Alabama Journal of Medical Science* (April 1986); Gina Kolata, *Flu: The Story of the Great Influenza Pandemic of 1918 and the Search for the Virus That Caused It* (1999); Ann McLaurin, *Journal of the North Louisiana Historical Association* (January 1982); Mississippi State Board of Health, *Report of the Board of Health of Mississippi from July 1, 1917 to June 30, 1919* (1919); Mississippi State Medical Association, *Transactions of the Mississippi State Medical Association* (1919); Dorothy A. Pettit and Janice Bailie, *A Cruel Wind: Pandemic Flu in America, 1918–1920* (2008); John B. Thomison, *Tennessee Medicine: Journal of the Tennessee Medical Association* (August 2006); United States Department of Health and Human Services, *The Great Pandemic: The United States in 1918–1919* (October 2011).

LeConte, John and Joseph

(JOHN, 1818–1891; JOSEPH, 1823–1901) SCIENTISTS.

Natives of Liberty County, Ga., John LeConte and his brother, Joseph, were sons of Lewis and Ann Quarterman LeConte. Descended from a French Huguenot, Lewis operated a large plantation and became an able amateur scientist. Both John and Joseph graduated from the University of Georgia and earned the M.D. degree from the College of Physicians and Surgeons in New York. After the death of their father, each inherited land and slaves. The brothers operated their plantations through overseers until the end of the Civil War. John practiced medicine in Savannah, Ga., from 1843 until 1846, when he was appointed professor of physics and chemistry at the University of Georgia. Joseph established a medical practice in Macon, Ga., in 1847, but three years later he abandoned it in order to study under Louis Agassiz at the Lawrence Scientific School of Harvard University. Upon completing his studies in 1851, he returned to Georgia and accepted the professorship of science at Oglethorpe University. In 1853 he was appointed professor of natural history at the University of Georgia.

As a result of a dispute with the university president, John resigned in 1855, and Joseph left for the same reason a year later. After serving for one year as a lecturer at the College of Physicians and Surgeons, John accepted a post at South Carolina College in 1857, where he was joined by his brother at the same time. Held in high regard, the LeConte brothers gained national recognition as scientists during the antebellum period, but they were severely set back by the Civil War.

Devoted defenders of the South, the LeConte brothers deplored the views of the Radical Republicans, and in 1869 they accepted posts at the newly established University of California, where John served as acting president in the 1869–70 academic year and as president from 1875 to 1881. At the time of his death, John was a member of several scientific organizations and had published more than 80 articles in medicine and physics. Joseph continued to write and eventually published more than 190 articles and nine books. A universalist, he wrote on numerous topics, including education, philosophy, religion, evolution, geology, and physiological optics. His books on vision, geology, and evolution and religion received international notice. Both LeContes were elected to membership in the National Academy of Sciences, and Joseph later served as president of the American Association for the Advancement of Science (1892) and the Geological Society of America (1896). A devoted camper, he was a charter member of the Sierra Club.

LESTER D. STEPHENS
University of Georgia

Emma LeConte, *When the World Ended: The Diary of Emma LeConte*, ed. Earl Schenck Miers (1987); Joseph LeConte, *Autobiography*, ed. William Dallam Armes (1903); John Samuel Lupold, "From Physician to Physicist: The Scientific Career of John LeConte, 1818–1891" (Ph.D. dissertation, University of South Carolina, 1970); Lester D. Stephens, *Joseph LeConte: Gentle Prophet of Evolution* (1982).

Leprosy

Persons afflicted with leprosy consider the term "leper" an odious label. Many even oppose use of "leprosy," preferring

"Hansen's disease," or HD, named after Gerhard Henrik Armauer Hansen, the Norwegian who discovered the bacilli. In Cajun culture, leprosy was *la maladie que tu nommes pas* (the disease you do not name). Patients today generally agree that the stigma of leprosy is worse than the disease.

The perception that leprosy is "a disease of the soul" is deeply rooted in the Western psyche. Stereotypes like gross deformities, fierce contagion, and body parts falling off are commonplace. In fact, leprosy is probably the least contagious of the communicable diseases. More than 95 percent of humans are naturally resistant. If left untreated, however, symptoms include nerve damage; loss of sensation in the arms, legs, and face; skin lesions; blindness; and, eventually, shortening of the nose, fingers, and toes (hence the belief about parts falling off). In cases where arms and legs are anesthetized by nerve damage, sores, blisters, and infections often go unrecognized, leading to amputations that further the myths.

Despite being the oldest disease known to man, much about leprosy remains unknown. There is no vaccine for leprosy. There is no test to determine susceptibility. The bacillus (*M. leprae*) has yet to be cultivated in vitro. Scientists theorize the disease is spread through the respiratory tract, but the exact mode of transmission is unknown. The period of incubation after exposure can range from 3 to 20 years. Men account for nearly 70 percent of all leprosy patients. The disease affects all races except Native American.

The history of leprosy treatment in the South is long and complex. In 1894 panic swept through New Orleans, La., after a report in the *Daily Picayune* revealed that seven "lepers" occupied a pesthouse inside the city limits. Before mobs acted on threats to burn the house, a compassionate physician at Tulane Medical School, Dr. Isadore Dyer, convinced the Louisiana legislature to lease an abandoned plantation 15 miles south of Baton Rouge on an isolated drop of land formed by a bend in the Mississippi River. Area residents were told the land would be used for an ostrich farm. On the evening of 30 November 1894, the seven leprosy patients were transported upriver on a coal barge. Conditions at their new home were deplorable—no running water, dilapidated buildings, and no regular medical care.

In 1896 four nuns from the Order of the Daughters of Charity of St. Vincent DePaul arrived at the home to take up duties as caretakers of the residents. The sisters shared quarters with rats and snakes, but also worked to improve the physical and spiritual health of the residents. According to the "inmate" reports, the sisters instituted prisonlike rules, including segregation of the sexes. For 27 years, the Louisiana Leper Home kept victims of the disease out of reach and out of sight. Gov. William W. Heard pleaded with Louisiana legislators, in the interest of economics, to deny the home's existence.

Despite pleas from leprologists and public health officials during the early 20th century, the U.S. Congress was reluctant to establish a national leprosarium until 1914, when John Early, a North Carolina native who contracted the disease during the Spanish-American War, made an appearance in Washington, D.C. Early had endured nearly a decade of makeshift exile, shuffled among tents on the banks of the Potomac, filthy boxcars, and abandoned houses. Despite his condition, Early had a penchant for publicity. He checked into the fashionable Willard Hotel in Washington, D.C., where senators, representatives, and the vice president were fellow guests, and called reporters from the *Washington Times* to demonstrate the ease with which the 400 "lepers at large" could mingle with ordinary citizens. Early's wish for a permanent home was eventually granted, in 1917, when Congress authorized $250,000 for the establishment of a national leprosarium. But with Early's victory came an unintended consequence: a national policy mandating quarantine for all who contracted leprosy.

In 1921 the U.S. Public Health Service took control of the Louisiana Leper Home. Dr. Oswald Denney, the former director of the leper colony at Culion in the Philippines, was appointed chief medical officer of U.S. Marine Hospital No. 66 (though the facility was called "Carville" owing to its proximity to the tiny Louisiana community). Upon arrival, Denney raised the yellow jack flag, an emblem of quarantine, next to the U.S. flag, which stirred the ire of the patient population. But Denney also brought innovation. He eliminated segregation of the sexes, sponsored the first patient dance, and embarked on major renovations to the facility, including 2.5 miles of covered walkways (sunlight, it was believed, aggravated leprosy).

Despite improvements, patients suffered great hardship for the next two decades. Newly diagnosed patients were often transported in shackles, their families were torn apart, and personal belongings were incinerated. Upon entering the leprosarium, patients took on aliases to protect family members from prejudice and shame. Patients were denied the right to vote. Outgoing mail was sterilized. The local Coca-Cola bottler refused to accept returnable bottles from the colony (patients used them to line flower beds). Marriage was prohibited. If a patient did become pregnant, the newborn was immediately taken away by the nuns and placed with a foster family. A cure for leprosy was elusive, but the patients lined up for experimental treatments, including injections of Chaulmoogra oil, ingestion of tree bark, and immersion in fever machines.

A unique culture developed within the walls of the colony. The patients formed their own Mardi Gras crews, sports fans regularly absconded through a hole in the fence to attend LSU football games, and a group of young patients joined the local softball league, but all games were played at home. Carville became a self-sufficient society with its own power plant, library, post office, churches, and a canteen. Entrepreneurship blossomed. Patients became barbers, cabinetmakers, repairmen, even bartenders.

The 1940s and 1950s brought scientific advancement, as well as social progress within the leprosarium. New medicine (sulfone drugs) showed promise in stopping the progression of the disease and rendering patients noncontagious. At the same time, the patients became their own strongest advocates.

Sidney Levyson, a young Jewish pharmacist from Texas, contracted leprosy at the age of 19. Upon his arrival at Carville, in 1931, he took an alias: Stanley Stein. In 1933 Stein launched the *Star*, a patient-run magazine. Originally published as a sort of town crier for the patients, the magazine soon became an important voice for the isolated community. The publication motto, "Radiating the Light of Truth on Hansen's Disease," accurately reflected Stein's passion. He fought to end the stigma associated with the disease. However, Stein struggled to find a balance between his desire to eradicate the terms "leprosy" and "leper" and his commitment as a journalist to freedom of press and speech. The masthead of the *Star* read: "We dislike the word 'leprosy' intensely, but we dislike the practice of censorship even more."

Stein inspired others at the leprosarium. The residents formed a patient federation, embarked on letter-writing campaigns to Congress, and fought for the right to vote and marry. Stein gradually lost his sight to leprosy, but he continued to edit the *Star* into the 1960s when its readership reached 50,000 worldwide. Funk & Wagnalls published his memoir, *Alone No Longer*, in 1963. Stein died in 1967, but the *Star* continues "The Carville Crusader's" vision to fight stigma associated with Hansen's disease.

Along with medical advancements came freedom. In the late 1950s, the gates of the national leprosarium were unlocked and 298 patients were set free. One year later, 276 remained at the colony. What once had been a prison for patients had now become a refuge from society. After much debate, the federal government agreed to allow anyone who had been forcibly quarantined to return to the facility to live out their lives. Dr. Edgar Johnwick, medical officer in charge of the leprosarium, proclaimed, "No one should be discharged against his will. No one should be kept against his will." Patients are buried in a cemetery at the back of the colony. Some of the tombstones are engraved with aliases.

In the 1960s and 1970s, the facility developed into the world's premier center of leprosy research. In 1971 scientists discovered that leprosy could be cultivated in the belly of the nine-banded armadillo. With that discovery, the era of patients as guinea pigs came to an end. A multidrug therapy was developed to battle resistant strains of the disease. And thalidomide was identified as an effective treatment for life-threatening reactions to multidrug therapy.

As the Carville patient population dwindled, budget constraints led to discussions of closing the facility. Louisiana congressman Gillis W. Long (cousin to Huey Long) secured temporary funding. The leprosarium was later named for him. Several attempts to subsidize the center's $17 million annual budget failed, including a two-year experiment when federal prisoners shared the facility with leprosy patients.

In 1999 the Gillis W. Long Hansen's Disease Center was sold to the Louisiana National Guard, where it established a boot camp for Louisiana's juvenile offenders. Carville residents were relocated to the third floor of Summit Hospital in Baton Rouge. Approximately 30 long-term leprosy patients refused to leave. About a dozen men and women reside in an isolated corner of the place they still call home.

Today, approximately 5,000 cases of leprosy exist in the United States; about 3,000 require treatment. On average, 150 new cases are diagnosed each year. The majority of new cases appear in immigrant populations. Of the indigenous cases, virtually all are discovered in south Louisiana and south Texas residents where there is a high prevalence of armadillo infected with the disease.

In 2011 leprologists discovered similar strains of *M. leprae* in armadillos and in humans—strengthening the theory (and what was thought of as an old wives tale) that the disease can be passed from animals to humans. Whole-genome resequencing of *M. leprae* from one wild armadillo and three U.S. patients with leprosy revealed that the infective strains were essentially identical.

NEIL WHITE
Oxford, Mississippi

Julia Rivera Elwood, ed., *Known Simply to the Rest of the World as Carville* (1994); Marcia Gaudet, *Carville: Remembering Leprosy in America* (2004); Tony Gould, *A Disease Apart: Leprosy in the Modern World* (2005); Zachary Gusssow, *Leprosy, Racism, and Public Policy* (1989); Michelle T. Moran, *Colonizing Leprosy: Imperialism and the Politics of Public Health in the United States* (2007); R. W. Truman et al., *New England Journal of Medicine* (28 April 2011); Neil White, *In the Sanctuary of Outcasts: A Memoir* (2009).

Lewis, Henry Clay
(1825–1850) PHYSICIAN AND HUMORIST.

Like several other newspaper humorists of the Old Southwest—Francis James Robinson, Orlando Benedict Mayer, and Marcus Lafayette Byrn—Henry Clay Lewis was a frontier doctor and a writer by avocation. Born in Charleston, S.C., on 26 June 1825, Lewis, who graduated from the Louisville Medical Institute in 1846, practiced medicine, first in Yazoo City, Miss., and then in rural Madison Parish on the Tensas River in northeastern Louisiana, and in late 1848 in Richmond, La., the county seat of Madison Parish. Many of Lewis's patients were planters and their slaves, farmers, swampers, and hunters—the real-life inspiration for some of the characters he featured in his humorous sketches.

Lewis's reputation as a humorist derives from a few sketches that, beginning in 1845 with "Cupping on the Sternum," began appearing in William T. Porter's weekly New York sporting paper, the *Spirit of the Times*, and a single book, *Odd Leaves from the Life of a Louisiana "Swamp Doctor"* (1850), which he published pseudonymously as Madison Tensas, M.D., and which consists of 22 sketches, many of them partly autobiographical and drawing on his experiences as a medical apprentice, medical student, and backwoods physician. The persona Madison Tensas, an old swamp doctor, recounts retrospectively his experiences, including some self-deprecating ones that make him look foolish, immature, and even reprehensible and cruel.

Lewis's humor, often dark and sinister, focuses on what Hennig Cohen and William B. Dillingham call the "underside of comedy," the grotesque and shockingly disgusting, the outlandish and horrifying, sometimes reminiscent of the grim subject matter of Edgar Allan Poe. Sketches in *Odd Leaves* such as "Valerian and the Panther," depicting a young boy having three of his fingers nearly bitten off when he tries to steal a plug from his sleeping father's mouth, and "The Indefatigable Bear Hunter," portraying a backwoodsman whose leg had to be amputated after a confrontation with a bear and the leg's wooden replacement, which the hunter subsequently uses as a weapon to bludgeon another bear, represent Lewis's brand of realistic sardonic humor.

Lewis's anticonventional and transgressive mode of humor, which sometimes defies limits and boundaries, going well beyond conventions of the acceptable and proper, is exemplified in his depiction of women and African American slaves who in selected sketches transcend the narrow boundaries of gender and race portraiture. In "A Curious Widow," Lewis destabilizes male dominance and misogyny by having a practical and cruel joke instigated by Tensas and some of his fellow medical students in which they sew the deformed and horrifying face of an albino cadaver to an oil cloth and place it in their room for the widow to find; but the prank intended to terrify her goes awry. When she expectedly finds the package containing the face, she laughs hysterically and verbally berates them into humiliation. Several of Lewis's sketches exhibit human dimensions of African American slaves and thereby counter and challenge racial stereotypes that emphasize black inferiority.

A celebrator of the unconventional, grotesque, and unsavory, Henry Clay Lewis, as Richard Boyd Hauck has accurately noted, is a "native absurdist who thrived upon the absurd ambivalence of the comic-horrible life," which makes him one of the most complex and intriguing of the South's antebellum humorists. Equally important, Lewis is a precursor of the brand of comic grotesquery prevalent in modern southern literature in selected works of Erskine Caldwell, William Faulkner, Flannery O'Connor, Harry Crews, Barry Hannah, Dorothy Allison, and others.

ED PIACENTINO
High Point University

John Q. Anderson, *Louisiana Swamp Doctor: The Life and Writings of Henry Clay Lewis* (1962); Hennig Cohen and William B. Dillingham, eds., *The Humor of the Old Southwest* (1994); Richard Boyd Hauck, *A Cheerful Nihilism: Confidence and "The Absurd" in American Humorous Fiction* (1971); Gretchen Martin, *Southern Literary*

Journal (Spring 2005); Ed Piacentino, in *Humor of the Old South*, ed. M. Thomas Inge and Edward J. Piacentino (2001).

Long, Crawford W.

(1815–1878) PHYSICIAN.

A general practitioner in the village of Jefferson, Ga., Crawford Williamson Long in March 1842 first used ether to anesthetize a patient, James Venable, before the removal of an encysted tumor from the back of his neck. He was thus one of the earliest southern physicians to make a major contribution to medicine.

Born in Danielsville and raised in Jefferson, Long graduated from Franklin College (University of Georgia). After reading medicine with a Jefferson doctor, Long studied at Transylvania and then transferred to the University of Pennsylvania medical school, receiving his M.D. degree in 1839. He gained surgical experience in New York hospitals before returning to Jefferson in 1841. Long thus acquired as sound a medical education as America could offer.

Late in 1841 Jefferson's young set, hearing of antics brought on by nitrous oxide administered to volunteers by an itinerant showman then crisscrossing the country, besought Long to make this gas for their own use. As a student in Philadelphia, Long had seen a showman use sulfuric ether, after which he had joined fellow students in private ether parties. Long superintended such entertainments for the pleasure of his Jefferson friends. His observations led him to use ether as an anesthetic.

The operation, with witnesses present, proved successful. Five times more Long used ether for surgery before publishing his results. In one case, Long amputated two injured fingers of a slave boy, employing ether for one operation but not the other. Long explained his delay in publishing by saying he wanted to prove that ether and not the impact of imagination negated pain. Long also had heard a Philadelphia professor condemn premature publication based on isolated experiments.

Long's article in the *Southern Medical and Surgical Journal* (1849) asserted his priority in using anesthesia against the quickly published claims of Harvard surgeon John Collins Warren and Boston dentist William T. G. Morton, who administered Letheon in 1846. Later, Long also competed with New England claimants for recognition and possible recompense from Congress for discovering anesthesia. Confusing claims and sectional tension kept such a bill from becoming law.

Anesthesia is a classic example of multiple discovery. Long's tardy publication hurt his claim to priority, but he never lacked southern defenders, including J. Marion Sims and his own apprentice, Joseph Jacobs, whose efforts eventually placed Long's statue in the national Capitol. Long made no further discoveries, practicing medicine and operating a pharmacy in Athens for the last 28 years of his life.

The Crawford W. Long Museum in Jefferson, Ga., honors Crawford's life and career.

JAMES HARVEY YOUNG
Emory University

Frank K. Boland, *The First Anesthetic: The Story of Crawford Long* (1950); Crawford W. Long, *Southern Medical and Surgical Journal* (December 1849); Ruby Lorraine Radford, *Prelude to Fame: Crawford Long's Discovery of Anaesthesia* (1969); Frances Long Taylor, *Crawford W. Long and the Discovery of Ether Anesthesia* (1928); James Harvey Young, *Bulletin of the New York Academy of Medicine* (March 1974).

Malaria

Malaria plagued the United States from colonial times until its practical eradication in the 1940s. Prevalent throughout the South and parts of the Midwest, it encompassed four diseases caused by the Plasmodium parasite: *falciparum*, *vivax*, *ovale*, and *malariae*. In America, it was manifest only in *vivax*, a mild form, and *falciparum*, a more deadly strain that could prove fatal by blocking the blood vessels. The parasites are transmitted to humans by the bite of an infected female *Anopheles* mosquito; after an incubation of 7 to 14 days, the disease produces very high fever, splitting headache, parched throat, and severe chills, followed by profuse sweating. An attack usually lasts 8 to 14 hours with episodes every 24, 48, or 72 hours, generating the labels "quotidian," "tertian," or "quartan."

Both *vivax* and *falciparum* parasites can lie dormant in the bloodstream for months, even years, and once acquired malaria recurs naturally. The *vivax* parasite, common in Europe, was transmitted to the New World in 1700. The *falciparum* parasite was rife along the West African coast and brought to the New World in the 1680s with the importation of slaves. Africans seemed to enjoy a degree of immunity from these fevers, leading some doctors to believe that they possessed an innate characteristic that made them an ideal tropical work force, an idea that supported scientific racism in the 19th century. But only some Africans carried the Duffy negative factor, which gives natural immunity to *vivax* malaria while another 10 to 20 percent carried the sickle cell trait, which offers a high resistance to *falciparum*. African Americans were as susceptible to malaria as others, particularly if they encountered new strains. By the early 20th century, southern physicians, influenced by the work of Robert Koch, agreed that blacks resisted malaria because repeated attacks in childhood produced immunity rather than having unique genetic factors.

People, plants, and animals in poorly drained lowland areas provided the perfect breeding ground for disease-carrying mosquitoes, which had a substantial effect on labor and settlement patterns in the American colonies. Settlers learned to drain their land and settle on high ground. By 1800 malaria was declining in the Northeast and West but increasing in the South, where mosquitoes survived year round in the warm climate. During the Civil War, the movement of large numbers of troops introduced parasites and vectors into areas formerly free of them. Occurring less frequently than only diarrhea and dysentery, malaria, commonly referred to as "intermittent fever," was the most reported disease. The Union army recorded nearly 1 million cases among white soldiers and almost 80,000 cases among blacks. Confederate forces had a higher incidence of malaria than the Union armies, but because they had some acquired immunity, it proved less fatal. Most physicians held a miasmatic theory of disease—that the fevers were caused by noxious vapors from putrefying animal or vegetable matter. Military officers were told to avoid poisonous miasmic wet areas and set up camps in dry elevations. Though based on flawed analysis, this directive kept troops away from the breeding grounds of *Anopheles*.

As a treatment and preventive, physicians used quinine—a drug comprised of powdered cinchona tree bark—with reasonable success. Soldiers returned home from the war with malaria parasites in their blood and countless epidemics spread in most southern states, producing a renewed awareness of the dangers. With an agriculture based on intensive hand-labor crops, such as cotton, workers were generally clustered in poorly constructed shanties located in highly malarious areas. The impoverishment of the South intensified the problem because many people lacked access to proper medical care and quinine. Virginia, the Carolinas, northern Florida, Alabama, Mississippi, and the Yazoo River area were particularly hard hit.

In 1880 the French military doctor Alphonse Laveran isolated the causal organism of the *Plasmodium* parasite. By the 1890s American scientists accepted the theory that mosquitoes became infected with *Plasmodium* by feeding on the blood of infected people. This led to new public health strategies to manage the disease, which was now officially called malaria. The decline of malaria in southern cities and towns began after the First World War and was largely the result of mosquito control campaigns: clearing and draining land, antilarval methods, and quinine therapy. With the socioeconomic turmoil of the Great Depression, there was a surge in malaria cases, but the situation improved in 1933 with Franklin D. Roosevelt's New Deal programs. Government agencies, including state and local governments, invested in mosquito eradication programs, but it was not until the prosperity following World War II that drainage, screening, pyrethrum sprays, mechanized farming equipment, and rural depopulation almost completely eradicated malaria in the South. The most effective solution was DDT (chlorophenothane), a new powerful chemical that was sprayed in areas where mosquitoes bred. The final elimination of malaria was a result of improvements in the southern economy and public health campaigns. The failure of the South to control malaria until well into the 1940s contributed to the idea that it was a backward region, but it also drew southerners together, formed an important part of regional identity, and contributed to the idea of "southern distinctiveness."

SHAUNA DEVINE
Duke University

Erwin Ackerknecht, *Malaria in the Upper Mississippi Valley, 1760–1900* (1945); Margaret Humphreys, *Malaria: Poverty, Race, and Public Health in the United States* (2001).

Maury, Matthew Fontaine

(1806–1873) CONFEDERATE GENERAL AND OCEANOGRAPHER.

Born 14 January 1806 near Fredericksburg, Va., and reared on a plantation at Franklin, Tenn., Matthew Fontaine Maury was appointed a U.S. midshipman in 1825. He sailed on three cruises, one being around the world, and rose to the rank of lieutenant before being permanently disabled in a stagecoach accident in 1839.

Maury attended Harpeth Academy, and his mastery of mathematics through calculus is evidenced in *A New Theoretical and Practical Treatise on Navigation*, published in 1836 and adopted by the navy for the instruction of midshipmen. In 1842 he was appointed superintendent of the navy's Depot of Charts and Instruments, and later, upon the completion of the U.S. Naval Observatory in Washington in 1844, he became its head as well.

Lieutenant Maury laid down an agenda for his new bureau and named the Observatory and Hydrographical Office, which called for an astronomical survey of the southern heavens and a scientific study of the twin oceans of water and air. Specifically, he proposed a systematic analysis of wind and current patterns by ships to find "tracks," or the most seaworthy routes. His "abstract logs" required the collection of empirical information about 22 different meteorological and oceanographic conditions, such as direction and rate of current, barometric readings, temperature of air and water, and the nature of winds and weather. As Maury explained in the *Southern Literary Messenger* in 1843, "Every new fact, however trifling it may seem, that is gathered from nature or her works, is a clue placed in our hands, which assists to guide us into her labyrinth of knowledge." After an international agreement at Brussels in 1853, more than 124,000 vessels, constituting over 95 percent of the world's shipping, cooperated in Maury's scientific undertaking. As millions of daily logs came to Maury's office, his findings were published in the *Wind and Current Charts* with their accompanying explanations and sailing instructions. As a result, clipper voyages to Australia, Brazil, and California were shortened by a quarter.

Maury early turned his research toward problems of interest to southern mariners and farmers, including the Isthmian canal, the Tehuantepec railroad, Mississippi River floodlands, and hurricanes. In attendance at various southern commercial conventions, he supported the establishment of direct trade with Europe through the use of steam packets. Because of adverse winds and the Gulf Stream, however, he found that sailing vessels to the south Atlantic ports made only 133 miles a day whereas northern ones averaged 162 miles. Nevertheless, he laid out a track for Savannah's vessels that required 39 days for European passage, which was a saving of 20 days over normal passage but still longer than New York's.

Although Maury pursued practical problems of weather and navigation, he also probed the basic mechanics of the sphere. His investigations of the dynamics of the Gulf Stream emphasized the effect of temperature and demolished John Herschel's theories on the trade winds. Maury was the first to chart the Atlantic seabed and to bring up samples of soil for microscopic examination of life forms. This work facilitated the laying of the first transatlantic cable in 1858. In the area of meteorology, he concluded that heated air currents and trade winds created hurricanes, and he projected such phenomena as the jet stream.

In 1855 Maury published the first modern work on oceanography, *The Physical Geography of the Sea*, with chapters on the Gulf Stream and Antarctic climatology, among others. Although not all contemporaries accepted his bold interpretations, this book was translated into most European languages and appeared in 20 English editions alone before the turn of the century. After the Civil War, in which Maury served the Confederacy as a technical expert on submarine mines and as a naval purchasing agent in Europe, he continued his writing. He spoke often on science and religion, with emphasis on the earth and its systems as a giant mechanism. In 1867 Maury made a profession of faith and joined the Protestant Episcopal Church. He died in 1873, while teaching at Virginia Military Institute.

HAROLD S. WILSON
Old Dominion University

Steven J. Dick, *Sky and Ocean Joined—The U.S. Naval Observatory, 1830–2000* (2003); Chester G. Hearn, *Tracks in the Sea: Matthew Fontaine Maury and the Mapping of the Oceans* (2002); Matthew Fontaine Maury, *Explanation and Sailing Directions to Accompany the Wind and Current Charts* (1858–59), Papers, Library of Congress, Washington, D.C., *The Physical Geography of the Sea* (1855), *Southern Literary Messenger* (August 1843); Frances Leigh Williams, *Matthew Fontaine Maury: Scientist of the Sea* (1963).

McDowell, Ephraim

(1771–1830) PHYSICIAN.

Of Scots-Irish descent, Ephraim McDowell was born in Rockbridge County, Va., the ninth of 11 children of Samuel and Mary (McClung) McDowell. In 1784 the elder McDowell, a former Revolutionary army officer and member of the Virginia legislature, moved his family to the small community of Danville in the Kentucky district where he served as land commissioner and magistrate. After completing his preliminary education, young Ephraim was apprenticed to Dr. Alexander Humphreys of Staunton, Va., an eminent physician and teacher. Then in 1793 and 1794 he attended medical lectures in Edinburgh, where he was influenced especially by the famous Scottish surgeon-anatomist John Bell. Returning to Danville in 1795 without a degree (the University of Maryland awarded him an honorary M.D. in 1823), McDowell soon established an extensive practice in the surrounding area. In 1802 the successful young doctor married Sarah Shelby, the daughter of Kentucky's first governor.

At a time before the development of anesthesia and detailed knowledge of the causes of infection, the name of Ephraim McDowell is associated with that of Jane Todd Crawford in one of the most celebrated cases in the annals of surgery. In December 1809 McDowell was called to a village 60 miles from Danville to consult with physicians whose 47-year-old patient appeared to present a complicated pregnancy. Upon examining Jane Todd Crawford, he determined she was not pregnant and that the swelling in her abdomen was a huge ovarian tumor. He advised her of the gravity of the circumstances, explaining that surgery was unprecedented and likely to be fatal; he offered to operate if she would come to Danville. On Christmas Day 1809, assisted by his nephew, McDowell performed the first ovariotomy in his home, removing a 20-pound tumor in the 25-minute operation during which Crawford recited the Psalms. She recovered fully and lived to the age of 78. The medical profession initially received a delayed report of this first case and two others with incredulity and harsh criticism, but McDowell's reputation was subsequently vindicated. Ironically, this famous pioneer in abdominal surgery is believed to have died of acute appendicitis. The McDowell house and apothecary in Danville, a national historic site, are owned and maintained by the Kentucky Medical Institute.

JOHN H. ELLIS
Lehigh University

John Duffy, *The Healers: A History of American Medicine* (1979); Laman Gray Sr., *The Life and Times of Ephraim McDowell* (1987); James Thomas Flexner, *Doctors on Horseback: Pioneers of American Medicine* (1939); J. N. McCormack, ed., *Some of the Medical Pioneers of Kentucky* (1917).

Medical Committee for Human Rights

In the spring of 1964, Mississippi civil rights activists invited upward of a thousand volunteers, most of whom would be white northern college students, to come down during the summer to work with local people on a variety of projects. To meet the medical needs of this influx of volunteers, Jackson physician Robert Smith contacted friends in New York asking for their help. That original group of doctors, predominantly Jewish and leftist in their politics, formed the Medical Committee for Human Rights (MCHR) and recruited more than 100 health-care professionals—doctors, nurses, psychologists, and social workers—to spend part of their summer vacations in Mississippi. Although some of the founding physicians were in private practice, others taught in medical schools or were working in the field of public health. Just a few of the physicians who went south during Freedom Summer were black; all of the Mississippi doctors they worked with were black.

After arriving at the Jackson airport, each team met with Dr. Smith and with civil rights leaders who directed them to project locations across the state, where they performed a variety of functions. They examined summer volunteers, treating their minor ailments. They also discovered that some movement veterans were suffering from what the psychologist Robert Coles referred to as "battle fatigue." Providing psychiatric care and "rest and recreation" opportunities for burnt-out movement workers would become an important function of MCHR. Overall, the work of the Medical Committee during Freedom Summer was a success. The physicians did provide medical support, but more important was simply the fact that they were there, an adult professional presence that contributed a small sense of security to a besieged group of freedom fighters.

At summer's end, the health-care volunteers decided to make the Medical Committee for Human Rights a permanent organization, with a headquarters and paid staff in New York and chapters in major cities across the country. A predominantly white organization, the committee would continue to assist the civil rights movement in Mississippi and the rest of the South, but movement workers were now demanding that MCHR focus its energies on the black poor rather than on the young activists, who were generally in good physical health. With that in mind, MCHR hired a full-time director for its southern project, a young black psychiatrist named Alvin Poussaint, who supervised a team of three to five nurses, headed by Phyllis Cunningham and Josephine Disparti.

For the next two years, MCHR assisted civil rights workers and local blacks in Alabama and Louisiana in addition to Mississippi. Members participated in the 1965 Selma to Montgomery march, made a significant contribution to ending segregation in hospitals throughout the South, and picketed at the American Medical Association (AMA) conventions because the AMA permitted its southern affiliates to deny membership to black physicians. They also opened a free health clinic in the Mississippi hamlet of Mileston, which became the inspiration for MCHR's most lasting contribution, the creation of the comprehensive community health center. (A half century later 20 million needy Americans would receive their primary care at community health centers.) MCHR left the South in the fall of 1966 because of internal disputes, a financial crisis, and the rise of Black Power, with its implicit message that whites where no longer welcome in the movement.

In the late 1960s and early 1970s MCHR chapters in northern cities began to deal with health-care issues closer to home, as the organization became younger and more radical, identifying with the agenda of the New Left. MCHR activists opposed the Vietnam War, opened free clinics in inner cities (sometimes in cooperation with the Black Panthers), pressured medical schools to enroll more students of color, supported a woman's right to an abortion long before *Roe v. Wade*, and demanded quality health care for all Americans under a single-payer system. The failed campaign to enact national health insurance was one of the last major MCHR initiatives. The organization went into decline in the 1970s, beset by left sectarian battles in a country that would soon elect Ronald Reagan president.

The Medical Committee left its mark on American history and provided a model for organizations that succeeded it, such Physicians for Human Rights and Physicians for a National Health Program. Its most enduring legacy has been the continuing social activism of its former members. As one longtime activist observed, "We tried to do something then, but if there's anything lasting, it's us. It's the alumni of MCHR trying to keep the flame alive, not in MCHR's name anymore, but with the same principles."

JOHN DITTMER
DePauw University

John Dittmer, *The Good Doctors: The Medical Committee for Human Rights and the Struggle for Social Justice in Health Care* (2009).

Medical Museums

In the 19th century, most college-educated physicians learned about bodies and anatomy through dissection, but anatomical knowledge was also acquired through study at medical museums and the preparation of specimens for museum displays. As medical college circulars and announcements, physician testimony, and surviving photographs indicate, museums were central to 19th-century professional medical education and research, illustrating the abstractions of the lecture hall and anatomical textbooks with the aid of preserved specimens, as well as being active sites of knowledge production and transmission. Early in the 19th century the South had few medical schools, but by the late 1850s the number of colleges in the region had increased significantly and the majority of these schools had large museum collections of anatomical and pathological objects and artifacts.

In 1850 the Louisiana legislature awarded $25,000 to the Medical Department of the University of Louisiana for an Anatomical Museum. Dr. A. H. Cenas (professor of obstetrics) and Dr. A. J. Wedderburn (professor of anatomy) visited Europe, bringing back to New Orleans a large collection of models and preparations produced by the world's leading anatomical and pathological model makers—such as Felix Thibert, Louis Auzoux, and Joseph Towne. Similarly, in the summer of 1859, Dr. Josiah Clarke Nott made extensive travels across Europe in pursuit of anatomical models and medical equipment, for what he hoped would become the best teaching museum in the United States at the new medical school being built in his hometown of Mobile, Ala. College medical museums were filled by wide-ranging networks of anatomical and patho-anatomical exchange as well as materials obtained on international collecting expeditions. Physicians in a medical college's rural hinterland and at lower levels of the profession, such as students or apprentices, regularly provided specimens to higher-status physicians who were on college faculties or had connections to hospitals and museums. Southern medical journals indicate that a large portion of the human specimens deposited in the region's medical museums was harvested from enslaved subjects.

Commercial anatomical exhibits, sharing the same varied displays as their medical counterparts, also started to appear in the mid-19th-century South and sought to bring the science of anatomy to the wider public. Dr. Reentz's World Renowned Cabinet Museum of Physiology and Anatomy was a touring exhibition on display at Charleston's Meeting Street Apprentices Library in 1857, charging the audience 50 cents for admission and providing lectures aiming to improve the intellect and elevate morals through a detailed knowledge of normal and pathological anatomy. In the late 19th century, as public attitudes toward nudity and sexuality began to change, popular anatomy museums retreated underground and became the preserve of sideshows and carnivals.

Professional medical museums reached a peak of professional influence in the early 20th century, and one of the most celebrated was that assembled by Edmond Souchon at Tulane University—which was even included as a highlight in the itinerary for Motor Tour 3 (Audubon Park and Universities) in the WPA Guide to the Crescent City. Portions of Nott's original museum collection survived in two locations: the Alabama Museum of the Health Sciences (University of Alabama at Birmingham) and at the Mobile Medical Museum. Opened in 1969 as the Museum of Medical Science, the Health Museum in Houston, Tex., offered 21st-century visitors a technologically mediated update of the 19th-century popular anatomical museum, providing both health information and an education in anatomy, including an interactive "hands-on" walking tour through the human body.

STEPHEN C. KENNY
University of Liverpool

The Health Museum, www.mhms.org; Mobile Medical Museum, www.mobilemedicalmuseum.com; Michael Sappol, *A Traffic of Dead Bodies: Anatomy and Embodied Social Identity in Nineteenth-Century America* (2002).

Medicine Shows and Patent Medicines

Distilling the appeal of sundry forms of popular entertainment, medicine shows peddled lowbrow culture and bogus curatives to rural and urban audiences throughout the United States from the early 1800s to the mid-20th century. Ventriloquists, dime-museum sensations, fortunetellers, and circus animals sated public appetites for diversion and propitiated cautious consumers with free spectacle. Additionally, the shows frequently regaled customers with music, and, indeed, many celebrated southern artists, including Jimmie Rodgers, Louis Armstrong, Roy Acuff, T-Bone Walker, Gene Autry, Sonny Terry, and Charlie Poole, worked the medicine show circuit at one time or another.

Between performances, the master of ceremonies, or pitchman, plied spectators with grandiloquent oration, boasting the virtue of patent tonics and elixirs. The success of such exhibitions largely relied upon the health concerns and relative isolation of attendees. To be sure, medicine show "doctors" capitalized on the anxiety of populations for whom medical care remained remote or unreliable, and for whom disease was a constant concern. Though cure-alls rarely worked as advertised, pitchmen's nostrums advertised relief for a great breadth of afflictions, from tapeworms and lethargy to constipation and cancer.

As Ann Anderson notes in *Snake Oil, Hustlers, and Hambones*, medicine shows carried on a tradition established by peripatetic Italian mountebanks (which translates to "one who mounts on a bench") during the Renaissance who hawked quack medicines, often with the aid of entertainers, to an ingenuous peasantry. The shows also drew on a variety of 19th-century pastimes and typically featured content associated with minstrelsy, vaudeville, and religious sermon, including blackface comedians, coon songs, variety acts, and Christian melodrama. By and by, a standard cast of characters emerged on the scene, and personas such as the "doc" (the doctor, usually the pitchman), Indian chief, cowboy, and rube regularly appeared on medicine show stages.

Not surprisingly, the proliferation of medicine shows remained closely connected to developments affecting patent medicine trade in the United States. The popularity of British nostrums helped to create a patent medicine market in North America during the late 18th century. Eventually, revolutionary sentiment settled in, and trade sanctions with the mother country compelled colonists to abandon imports for native concoctions. Thereafter, the consumption of American-made patent medicines inclined steadily with the burgeoning number of proprietary remedies made available. The evolution of the newspaper industry also precipitated increased use of patent medicines. Daily papers flourished as the costs of production declined in the wake of technological advancements, allowing editors to devote more space to promotional material. As the reading public grew, advertisements propelled patent medicines into an age of mass consumption.

Nostrums commonly carried the potential to both damage and improve the health of their consumers. In addition to herbal components and vitamins, patent medicines occasionally included such noxious ingredients as turpentine and arsenic. Narcotics were also regularly incorporated, and addiction posed a genuine risk. Likewise, alcohol often composed a significant part of many nostrums, a boon to the drinking public during prohibition when spirits were illicit, but proprietary medicines retained legal and moral sanction. As James Harvey observes in *The Toadstool Millionaires*, "store ledgers" indicated that in "Mississippi, some families replenished their stock of alcoholic nostrums every day."

Of course, the South's major contribution to the patent medicine industry was Hadacol, the brainchild of Dudley LeBlanc, a former Louisiana senator and Cajun whose traveling medicine show line-ups included the likes of Bob Hope, George Burns, Carmen Miranda, and Hank Williams over the years. Hadacol, largely consisting of Vitamin B and alcohol, proved to be a multimillion-dollar enterprise for LeBlanc, who blazoned customer testimonials in advertisements to imbue the nostrum with a panacealike aura. Though the Hadacol Caravan carried out immensely popular tours in the South and Midwest during the early 1950s, LeBlanc's extravagant tastes landed the company in massive debt, forcing him to sell his enterprise in 1951.

As pitchmen found particularly receptive audiences among rural populations, the South (along with the Midwest) hosted more medicine shows

Advertisements for patented wonder drugs, South Carolina, 1938 (Marion Post Walcott, photographer, Library of Congress [LC-USF-34-50583-D], Washington, D.C.)

than other regions of the country. Doc McDonald's Indian Medicine Show played small towns in Oklahoma, east Texas, and Arkansas during the first decades of the 20th century. A genuine Choctaw medicine man, Jude "Doc" McDonald, helmed the show with the assistance of his family, pitching to farmers who regularly traded homegrown goods for his nostrums until the onset of the Great Depression. Bill Kerr, another longtime pitchman from North Carolina, broke into the entertainment industry during the first decade of the 20th century with a grotesque yet impressive feat: biting the heads off snakes. However, because the novelty of such demonstrations quickly waned, Bill was unable to sustain a regular audience for his snake act and ultimately found success as a medicine show "doc" in South Carolina. Deploying the image and allure of Native Americans on his nostrums and in his acts, Kerr exploited the widely held belief that Indians possessed extensive knowledge with respect to natural healing methods. Bill hawked Kerr's Indian Remedy year-round, trailing seasonal economic booms in South Carolina and other southern states. Though he used an all-white cast for a short time after the Ku Klux Klan intimidated Kerr's interracial crew, Bill employed venerated bluesman Pink Anderson in his medicine show for more than 30 years until Kerr quit the business in 1945.

Medicine show popularity reached its acme toward the end of the 19th century and began a slow decline in the first decades of the 20th century, sustaining a conjunctive blow from technological advancements in the entertainment industry and legislation regulating the content of both advertisements and drugs. As radios increased in popularity, allowing Americans to access entertainment from the comfort of their own homes, many medicine show acts and advertisements transitioned to the airwaves. Movies—America's most popular form of entertainment in the first half of the 20th century—likewise outcompeted medicine shows as a divertissement. By the mid-1930s, legislation placing harder drugs under the jurisdiction of pharmacies was widespread, and the Food, Drug, and Cosmetic Act of 1938 forced patent medicine manufacturers to fully disclose the ingredients of their remedies, quashing the ability of medicine shows to continue with business as usual. Heralding the end an era, the Great Depression hampered the livelihoods of both medicine shows and their patrons, and by midcentury they had all but disappeared. Nonetheless, the medicine show legacy remains evident in contemporary popular culture, as infomercials, advertisements, and commercials continue the gainful tradition of pairing entertainment and enterprise.

KATHRYN RADISHOFSKI
University of Mississippi

Ann Anderson, *Snake Oil, Hustlers, and Hambones: The American Medicine Show* (2000); Marshall Wyatt and Bengt Olsson, *Good For What Ails You: Music of the Medicine Shows, 1926–1937* (2005); James Harvey Young, *The Toadstool Millionaires: A Social History of Patent Medicines in America before Federal Regulation* (1961).

Meharry Medical College

The genesis story of what is now Meharry Medical College, the first medical institution in the South to train African American doctors, is anything but simple. The institution was founded by the Methodist Episcopal Church in 1876 as the medical department of Central Tennessee College, a school for freedmen chartered in 1867, but the story begins on the Civil War battlefield with two men on seemingly opposite sides eventually coming together for a greater good. According to a 1948 *Chicago Tribune* article by Lloyd Wendt, Confederate army surgeon Dr. William J. Sneed consistently crossed paths with Union Medical Corpsman George Whipple Hubbard as they worked together to mend the wounded. Hubbard, a Vermont native, expressed his desire to become a doctor one day if he survived the war, and Sneed urged him to come to Nashville. On the way to Nashville from Ohio where he had been demobilized, Hubbard's unnamed traveling comrade from Illinois became ill.

In Nashville, it looked like his friend would not survive, especially since neither traveler had any money. Sneed, who had also fallen on hard times, arranged for a former slave girl, who had been trained to care for a sick member of her master's family, to nurse the friend back to health with medicine supplied by Sneed. Sneed's traveling companion, along with Sneed, eventually helped financially support Hubbard through Nashville Medical College, which is now the Vanderbilt University Medical Department.

The excellent care provided by the unnamed black woman convinced the three men that other black women should be trained in the nursing profession. A visit to Central Tennessee College president Rev. John Braden, who was also white, resulted in a mutual decision to establish a medical school. But, although Sneed offered his services for free and Hubbard promised to join him, there was no money to fund it. Luckily the friend from Illinois remembered the story of the Rev. Samuel Meharry from Shawnee Mound, Ind.

As Meharry, a Scots-Irish immigrant salt trader, traveled through Kentucky, his wagon fell off the road into a swamp. He fell sick and was later found by a black farmer who took him to his humble home. There, Meharry was fed and nursed back to health. According to legend, when Meharry was able to recover his wagon, he regretted not having money to give the family but vowed to one day do something great for their race.

So, in 1875, Braden found Meharry, who, along with his four brothers, Alexander, David, Hugh, and Jesse, raised an initial $500 to establish a Department of Anatomy in the basement of Central Tennessee College. The following year, the brothers donated $40,000 toward a full medical college, with the first official class starting October 1876.

There was just one graduate from the first class in 1877 and three in the second class in 1878. In 1886 a dental department was established, with a pharmacy department following in 1889. Central Tennessee College underwent a name change in 1900, becoming Walden University in honor of Methodist bishop John Morgan Walden. Finally, in 1915, Meharry Medical College received its own charter, with Hubbard officially serving as the institution's first president until his death in 1921. Walden University moved to another campus.

Surviving its founding was half the battle. Meharry alum and one of the nation's first African American doctors, Dr. Robert F. Boyd, along with fellow Meharry alum Dr. Miles V. Lynk, founded what is now the National Medical Association, a professional organization connecting African American doctors, in 1895. The men recognized that a medical school could not survive without offering its students clinical experience. Because African Americans were banned from Tennessee hospitals, Boyd led the effort to create Meharry's own George W. Hubbard Hospital in 1910. As an added bonus, it also included a nurse-training department.

For much of the 20th century Meharry continued to grow. Under the leadership of its second president, Dr. John J. Mullowney, Meharry received an "A" rating from the Council on Medical Education and Hospitals in 1923 as well as moving to a larger location near Fisk University in 1931. From 1952 until his resignation (due to poor health) in 1966, Dr. Harold D. West, Meharry's first African American president and the first selected from its faculty, continued the expansion, leading a drive to raise $20 million. West also improved faculty salaries and benefits and added a new wing to the Hubbard Hospital.

At least 14 black medical schools were founded between 1865 and 1910, but only Meharry and Howard University Medical School survived, and from 1910 to 1970 these two schools graduated the majority of black physicians and dentists in the United States. Yet, in 1994, when Dr. John E. Mauphin, an alumnus of the dental school, took over for the previous president, Dr. David Satcher, who became the U.S. Surgeon General, Meharry was on the brink of closing.

Mauphin helped prevent the unimaginable by successfully completing the merger of Meharry's historic Hubbard Hospital, which had become one of the school's major financial drains, with the city of Nashville's Metro General Hospital. In addition, Mauphin cemented a controversial partnership with Vanderbilt University that helped attract much-needed research funding and modern-day prestige.

Meharry's 10th president, Dr. Wayne J. Riley, who succeeded Mauphin in 2007, once again raised Meharry's national profile when he became one of the advisors to President Obama's transition's team health-care working group.

Nearly 150 years after it was founded in 1876, Meharry remains true to its mission and prides itself on its ongoing commitment to eradicating health-care disparities in African American communities. Even in the 21st century, it still trains an estimated 20 to 25 percent of all African American physicians and doctors in the United States.

RONDA RACHA PENRICE
Atlanta, Georgia

Jessee Barber, *Journal of the National Medical Association* (April 1979); W. Michael Byrd and Linda A. Clayton, *An American Health Dilemma*, vol. 1: *A Medical History of African Americans and the Problems of Race, Beginnings to 1900* (2000); Axel C. Hansen, *Journal of the National Medical Association* (July 1973); David Hefner, *Black Issues in Higher Education* (July 2001); Todd L. Savitt, *Journal of the National Medical Association* (January 1996); T. Manuel Smith, *Journal of the National Medical Association* (November 1953).

Moore, Samuel Preston

(1813–1889) CONFEDERATE SURGEON.

Samuel Preston Moore, surgeon general of the Confederate army, is among the most poorly appreciated personages of the Civil War. Born in 1813 in Charleston, S.C., he received his early education in his native state and graduated from the Medical College of South Carolina 8 March 1834. One year later, he was commissioned assistant surgeon (with the rank of captain) in the U.S. Army, beginning a 26-year stint of service at military posts in various areas of the country. After serving in the Mexican War, he was made full surgeon in 1849. On 25 February 1861 Moore resigned from the army to avoid fighting against his native state. He entered medical practice in Little Rock, Ark., but on 30 July 1861 Confederate president Jefferson Davis appointed Moore acting surgeon general.

Moore faced an almost insurmountable task—the establishment of a medical department. Physicians, drugs, supplies, and hospitals had to be provided. Starting with only 24 physicians who had resigned from federal service, Moore during the course of the war recruited some 3,000 physicians for Confederate service. The medical department cared for 600,000 Confederate soldiers, 270,000 prisoners, and over 3 million wounded or sick persons. Moore soon established an extensive hospital system, ranging from general hospitals to convalescent facilities. He is credited with the introduction of hospital "huts," the forerunner of the pavilion hospital. To replace medical supplies blockaded by the enemy, Moore was responsible for the preparation of drug substitutes from indigenous plants. Moore required regular sick calls, sanitary inspections, and regular reports of all medical activities.

To advance training of his medical officers, Moore encouraged educational meetings, refresher courses, and the publication of practical manuals on military medicine. He was instrumental in organizing in August 1863 the Association of Army and Navy Surgeons of the Confederate States and in the publication of the *Confederate States Medical and Surgical Journal.*

From the outset Moore was a strict disciplinarian, and his rigid enforcement of regulations often seemed to men fresh from civilian life little short of tyranny. Complaints were not wanting, but it was widely agreed that the medical department was among the most efficient in the Confederacy. President Davis accorded the highest praise to Moore and his department.

After the war, Moore remained in Richmond and devoted most of his time to the advancement of precollege education and agriculture. His contributions in both fields were significant.

He died suddenly on 31 May 1889 and is buried in Hollywood cemetery in Richmond.

HARRIS D. RILEY JR.
University of Oklahoma Health Sciences Center

Bruce S. Allardice, *More Generals in Gray* (1995); H. R. McIlwane, *Surgery, Gynecology, and Obstetrics* (November 1924); P. N. Purcell, R. P. Hummel, *American Journal of Surgery* (October 1992).

Pellagra

Although pioneer nutritionist Casimir Funk believed that pellagra existed sporadically in America as early as 1880, Alabama reported the first epidemic during the summer of 1906, after which it soon spread rapidly across the South. Over the next five-year period there were nearly 16,000 cases with 39.1 percent mortality. The alarming death toll prompted action by the U.S. Public Health Service and Joseph Goldberger's indefatigable campaign to eradicate a disease characterized by the "four d's": dermatitis, diarrhea, dementia, and sometimes death. As early as 1914 Goldberger, observing pellagra in various institutional settings, concluded that it was a nutritional deficiency disease. Later, Goldberger, joined by North Carolinian George A. Wheeler and statistician Edgar Sydenstricker, made a series of field investigations into pel-

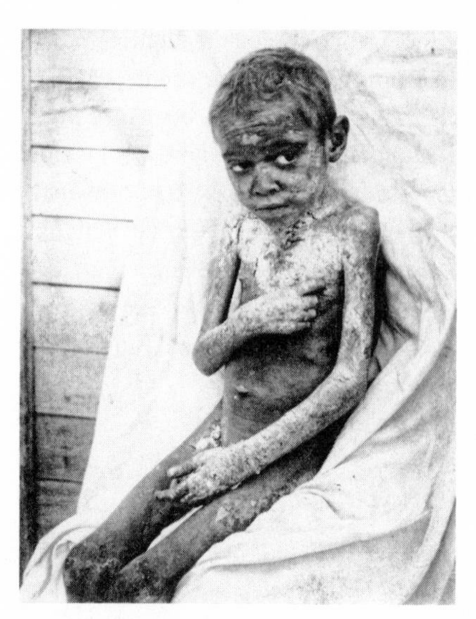

Young boy with pellagra, early 20th-century rural South (National Archives, Washington, D.C.)

lagra in seven cotton-mill villages in South Carolina. Their painstaking work yielded important conclusions about the nature of the disease, the most important being that pellagra was essentially a socioeconomic disease related to low family income and available food supplies. Goldberger's team revealed a disease not of the poor but of the *very* poor.

But Goldberger's thesis did not easily persuade colleagues or community leaders in the South. The truth was hampered by the Thompson-McFadden Commission's report in 1912 that proclaimed the disease infectious in nature. The consequences of this much-touted scientific cul-de-sac would be persistently recited by Goldberger's opponents like a well-worn catechism of faith because it told southern leaders (medical and otherwise) precisely what they wanted to hear: pellagra was *not* rooted in the socioeconomic conditions of the region but rather in the whimsy of microbes facilitated by bad habits, ignorance, and inattention to hygiene—the "lazy" disease. Meanwhile pellagra rates climbed, reaching 170,000 cases by 1927 and exceeding 200,000 through 1930. Although Goldberger would die in 1929 without finding the specific cause of the disease (that would await Conrad Elvehjem's studies in the 1930s, which conclusively linked the disease to niacin deficiency), his nutritional deficiency thesis was, fortunately, generally accepted by 1927 when the Southern Medical Association passed a resolution in its favor. The next year pellagra mortality would reach an all-time high (see Figure 2). Yet naysayers persisted. Typical was Alabama physician William Partlow, who insisted, "The late Doctor Goldberger did not, nor has any one else proven pellagra to be due to a dietary deficiency."

Why the resistance? Goldberger's team knew that for poor whites and blacks life in the "new" human chattel-free agrarian South transformed into a slavery of a different kind, a color-blind bondage of economic dependence upon tenant farming under an equally merciless white master—cotton. But if the old agrarian ways of Dixie offered only the prospect of a dismal life under new arrangements, the so-called New South of business and industry was showing itself to be equally bereft of worthwhile returns. Indeed the New South was built upon a bitter contradiction. Touting the benefits of modern diversified agriculture and industrial expansion, the New South stood upon the foundations

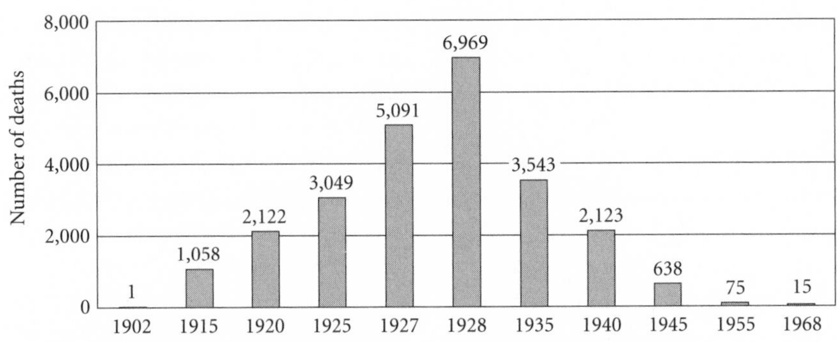

FIGURE 2. *Pellagra deaths in the United States*
Source: *U.S. Public Health Service Center for Disease Control, Atlanta, Ga.; and U.S. Vital Statistics, U.S. Bureau of the Census, Department of Commerce.*

of a low wage base managed by textile manufacturers and mining magnates quick to take advantage of the postbellum opportunities offered by the vanquished region. Southern boosters proclaiming an "abundant supply of labor, thrifty, industrious, and one hundred percent American," belied an income only 60 to 70 percent that of other Americans. Although tenant farmers were bound to their local stores and sundry shops through lines of credit that dropped them deeper and deeper into debt, factory and mine workers were frequently paid in company script redeemable only at the company store or at stores that accepted the company vouchers. This arrangement tied workers to the store inventory, which could be ample or limited, either way, and even in the face of the most abundant store inventories, low wages and mounting debt forced even the most frugal families away from choice meats and fresh fruits and vegetables and toward cornmeal and biscuits (since 1878 produced by a roller-mill process that reduced the meal and flour's protein, ash, and vitamin content by 80 to 90 percent), molasses, syrup, fatback, and coffee, a menu tailor-made for pellagra. A similar scenario played out for sharecroppers and tenant farmers who were tied to the credit lines of landowners. Low wages and high debt bred pellagra.

Unfortunately, the challenge of pellagra would not prompt wide-scale action to eradicate poverty in the South. Instead, it was food enrichment incident to federal mandates issued during World War II that would end the southern scourge, a point clearly demonstrated in the precipitous decline in mortality rates by war's end.

MICHAEL A. FLANNERY
University of Alabama at Birmingham

Elizabeth W. Etheridge, *The Butterfly Caste: A Social History of Pellagra in the South* (1972); Casimir Funk, *The Vitamines* (1922); Joseph Goldberger, *Goldberger on Pellagra*, ed. Milton Terris (1964); Alan M. Kraut, *Goldberger's War: The Life and Work of a Public Health Crusader* (2003); Youngmee K. Park, Christopher T. Sempos,

Curtis N. Barton, John E. Vanderveen, and Elizabeth A. Yetley, *American Journal of Public Health* (May 2000).

Poteat, William Louis

(1856–1938) WAKE FOREST COLLEGE PRESIDENT AND INTELLECTUAL.

William Louis Poteat embodied Progressive Era leadership in the American South. He was a teacher of science, a college president, a Baptist liberal, an ardent Prohibitionist, a participant in myriad reform campaigns, and an advocate of freedom of thought and inquiry. His parents, James and Julia Poteat, owned a large plantation and an enslaved labor force near Yanceyville in Caswell County, N.C. William, nicknamed Bud Loulie as a youngster and Dr. Billy later in life, initially enjoyed a privileged childhood, but the end of the Civil War led the family to become hotelkeepers in Yanceyville, with tenants working the plantation. William left the turmoil of Reconstruction when he enrolled at the North Carolina Baptist State Convention's Wake Forest College in 1872. The institution was a natural fit, as his father had held positions of local and regional importance in the denomination.

William Poteat made Wake Forest College his life's work. He graduated in 1877, became a tutor of languages there a year later, and was named a science professor in 1880. He largely taught himself the scientific knowledge he needed, though summer stints at the University of Berlin and the Marine Biological Laboratory at Wood's Hole, Mass., gave impetus to his relatively early adoption of laboratory work as a teaching tool. He dabbled in observation-based biology research with modest professional success, but he earned regional renown as a teacher of the subject.

By the close of the 1880s, Poteat had reconciled his Baptist religious upbringing with his growing knowledge of evolutionary biology, largely by deciding that science and religion operated in the separate spheres of reason and faith. Though his published views on the topic caused mild controversy at the turn of the century, the trustees of Wake Forest chose him as president of the college in 1905, a position he held until June 1927.

Just as the study of biology influenced his religious views, so too did his reading of the so-called higher criticism, which historicized biblical writings. Poteat moved away from a literalist understanding of the scriptures and toward emphasizing the duty of Christians to reform society. Although he did not fully adopt Social Gospel theology, Poteat did push fellow faithful in the South to mix their traditional evangelism with work for social improvement.

Poteat's greatest intellectual legacy is his willingness to defend the teaching of evolution during the 1920s. Facing intense criticism from supporters of conservative and fundamentalist Christianity, Poteat made a dramatic address at the December 1922 meeting of the Baptist State Convention. He forthrightly defended his religious faith and survived calls for his resignation as Wake Forest's president. Those calls continued through the mid-1920s, but Poteat nonetheless helped rally opposition to proposed state legislation that would have banned the teaching of evolution. In 1925 the University of

North Carolina, led by Harry W. Chase, selected Poteat to deliver the McNair Lectures at the Chapel Hill campus. He put forth his trademark ideas about the compatibility of religion and science, again raising a statewide furor. The controversy continued when the young University of North Carolina Press published the three talks as *Can a Man Be a Christian To-day?*

Poteat delayed his retirement as president until he felt freedom of intellectual inquiry was reasonably secure at Wake Forest College, and during his retirement he remained highly visible as a teacher, speaker, and Prohibitionist. In 1936 the Baptist State Convention chose him as its president, a fitting capstone to his long and varied career.

RANDAL L. HALL
Journal of Southern History
Rice University

George McLeod Bryan, "The Educational, Religious, and Social Thought of William Louis Poteat as Expressed in His Writings, Including Unpublished Notes and Addresses" (M.A. thesis, Wake Forest College, 1944); Randal L. Hall, *William Louis Poteat: A Leader of the Progressive-Era South* (2000); Suzanne Cameron Linder, *William Louis Poteat: Prophet of Progress* (1966).

Reed, Walter

(1851–1902) PHYSICIAN.

Major Walter Reed, who was one of the foremost bacteriologists and epidemiologists in the nation during the formative years of modern medicine, is best known for his work as chairman of the U.S. Yellow Fever Commission and discoverer of the mode of propagation of the disease.

Reed was born near Gloucester, Va., on 13 September 1851 and spent his childhood moving around the countryside of Virginia and North Carolina with his father, a Methodist minister. He received an M.D. degree from the University of Virginia in 1869 and went on to work for several years in New York hospitals. In 1874 he became a lieutenant in the U.S. Army Medical Corps. Though he was able to study pathology under William H. Welch at Johns Hopkins for a while, he spent most of his army career at isolated and dreary outposts.

He first began to demonstrate his skill as a medical investigator in Washington, D.C., during a virulent outbreak of malaria in 1896. When the Spanish-American War broke out in 1898, Reed was appointed to direct an investigation of typhoid. His findings made it possible to end the deadly epidemic of that disease in army camps.

In 1900 Reed became the head of an army board assigned to investigate the cause of yellow fever, which was rampant in Cuba at that time and also paid regular and deadly visits to the United States, especially in the Gulf States and the Lower Mississippi River Valley. Reed and his colleagues traveled to Cuba to search for the origins of the dread disease. Their experiments there led Reed to ascertain that yellow fever was carried by the *Aedes aegypti* mosquito. Once the cause was identified, steps were taken to eradicate the mosquito and thus the disease. The implications of Reed's work for the American South, where yellow fever took devastating tolls in life and productivity throughout the 19th century, were enormous.

Reed died following surgery for a ruptured appendix on 23 November 1902. The Walter Reed General Hospital, then the Walter Reed Army Medical Center (WRAMC), in Washington, D.C., opened in 1909 and was named in his honor. WRAMC closed in 2011 and was replaced with the Walter Reed National Military Medical Center, also in Washington.

LUCIE R. BRIDGFORTH
Memphis State University

Century Magazine (October 1903); Molly Caldwell Crosby, *The American Plague: The Untold Story of Yellow Fever, the Epidemic That Shaped Our History* (2006); *New York Times* (2 November 1902); John R. Pierce, *Yellow Jack: How Yellow Fever Ravaged America and Walter Reed Discovered Its Deadly Secrets* (2005); Albert E. Truby, *Memoir of Walter Reed: The Yellow Fever Episode* (1943); Laura N. Wood, *Walter Reed: Doctor in Uniform* (1943).

Research Triangle Park

Research Triangle Park (RTP) is a planned industrial research park in North Carolina. Its "Triangle" includes more than 7,000 acres and is formed from the geographic locations of three research universities: Duke University, North Carolina State University, and the University of North Carolina at Chapel Hill. Developed and managed by the nonprofit Research Triangle Foundation, the park contains industrial laboratories and trade associations, federal and state government laboratories, nonprofit research institutes, and university-related research organizations in 22.5 million square feet of developed space. The park's 170 research organizations employ 42,000 full-time workers, who receive a combined annual payroll of $2.7 billion.

North Carolina governor Luther Hodges initiated the Research Triangle program in 1955 with the appointment of the Governor's Research Triangle Committee of corporate and university leaders. With private funding, the governor's committee was incorporated in 1956, and sociologist George Lee Simpson Jr. from the Chapel Hill faculty was appointed director. The committee's plan was to promote the region for industrial research, and faculty members were initially employed to promote the idea. North Carolina's per capita income in the 1950s was one of the lowest in the nation ($1,049, compared to the national average of $1,639), so the committee hoped to attract industrial laboratories and high-technology industry, which would diversify the industrial base from the traditionally low-wage manufacturing industries such as tobacco, textiles, and furniture that dominated the region's economy. The Committee also hoped to reverse out-migration of North Carolina youth trained in science and engineering and to help the area's universities attract and retain science and engineering faculty members by expanded consulting opportunities. Although the committee began soliciting private investors for fundraising, its focus soon shifted to nonprofit ownership, and in December 1958 the committee became the Research Triangle Foundation, Inc., and the Research Triangle Institute and Park were established in January 1959.

Research Triangle Park is now the

largest research park in the world, housing more than 130 research facilities in 2007. Areas of concentration of RTP research generally reflect the research strengths of the triangle's universities and now include environmental sciences, life sciences, pharmaceutical and agricultural chemicals, biotechnology, microelectronics, and information and communications technology. Discoveries influenced by RTP scientists and researchers include the invention of the Universal Product Code (UPC), 3-D ultrasound technology, Astroturf, and Taxol, an anticancer drug named most important in cancer treatment by the National Cancer Institute. The park also functions as a nurturing place for startup companies; more than half of the companies in the park (56 percent) have less than 10 employees. Some of the large companies represented at RTP include IBM, Cisco Systems, Sony Ericsson, Biogen Idec, Nortel, and NetApp. Some federal government agencies also conduct research there, including the Environmental Protection Agency, the U.S. Forestry Service, and the National Institute of Environmental Health Sciences.

Educational support activities in the park include the North Carolina Board of Science and Technology, the Triangle Universities Computation Center, and the Triangle Universities Center for Advanced Studies, Inc. (TUCASI), which holds 120 acres in the park for joint activities of the three universities. On the campus of TUCASI are the National Humanities Center and the Microelectronics Center of North Carolina.

Research Triangle Park has been instrumental in the transformation of the state's economy. It has helped diversify industry to higher-wage "New Line" industries, which include chemicals, electronics, communications, engineering, and management services. The Park has also spurred the development and upgrade of all of the region's universities. Although RTP was developed without state appropriations, the state provided leadership, cooperation, and the support of the park's educational base. The success of the Triangle is a notable example of effective cooperation among state government, higher education, and the corporate community.

WILLIAM F. LITTLE
University of North Carolina at Chapel Hill

MARY AMELIA TAYLOR
University of Mississippi

Victor J. Danilov, *Industrial Research* (May 1971); W. B. Hamilton, *South Atlantic Quarterly* (Spring 1966); Luther H. Hodges, *Businessman in the Statehouse* (1962); A. N. Link, *A Generosity of Spirit: The Early History of the Research Triangle Park* (1995); Research Triangle Park of North Carolina, "RTP: Evolution and Renaissance" (2006), rtp.org (2007); Ruth Walker, *Christian Science Monitor* (15 June 1982); Louis Round Wilson, *Louis Round Wilson's Historical Sketches* (1976).

Ruffin, Edmund

(1794–1865) AGRICULTURAL REFORMER.

The preeminent scientific agriculturist of the Old South and a dedicated southern nationalist, Edmund Ruffin was born in Prince George County, Va., the son of a prosperous James

River planter. He attended the College of William and Mary, served briefly in the War of 1812, and then embarked upon a nearly half-century career as a gentleman farmer. Plagued initially by lands impoverished by two centuries of tobacco culture, he set out to improve them. By means of an elaborate series of experiments conducted over the course of 15 years on his Coggin's Point estate, Ruffin demonstrated the useful properties of marl, a shell-like deposit consisting primarily of clay mixed with calcium carbonate, which neutralized soil acidity and rendered sterile soils productive. He published his findings in 1832 under the title *An Essay on Calcareous Manures* and during the following decade spearheaded an agricultural renaissance in the Upper South through his distinguished journal, the *Farmers' Register*. In subsequent years he conducted an agricultural survey of South Carolina, proved the efficacy of his theories by converting his two Virginia farms, Beechwood and Marlbourne, into model estates, and served four terms as president of the Virginia State Agricultural Society.

Edmund Ruffin, agricultural reformer and southern nationalist from Virginia, near the end of his career (Virginia State Library and Archives [45.9232], Richmond)

Although his lasting fame derives from his contributions as an agricultural reformer, Ruffin is significant too as a representative of the planter elite of the slave South. Like others of that class, he exhibited a cultural and intellectual versatility. Thus, quite apart from his pragmatic interest in soil chemistry, he manifested a natural curiosity about scientific phenomena ranging from geology to ethnology. Cultured and well read—he had read all of Shakespeare's plays before he was 11 years old—he was in many respects a true renaissance man. Also typical was Ruffin's consuming interest in politics. Too opinionated and too much a party maverick to stand for elective office—he served only a partial term in the Virginia state senate—Ruffin, nevertheless, immersed himself completely in the secession movement following his retirement from farming. When his labors bore fruit in 1861, he was accorded the honor of firing the first shot at Fort Sumter. Four years later, broken in spirit and fortune by the demise of his beloved Confederacy, the embittered Ruffin, in a gesture that symbolized not only his personal tragedy but that of his region, took his own life rather than submit to the anticipated indignities of Reconstruction.

WILLIAM K. SCARBOROUGH
University of Southern Mississippi

David F. Allmendinger, *Ruffin: Family and Reform in the Old South* (1990); Avery O. Craven, *Edmund Ruffin, Southerner* (1932); David R. Detzer, *Allegiance: Fort Sumter, Charleston, and the Beginning of the Civil War* (2001); William M. Mathew, *Edmund Ruffin and the Crisis of Slavery in the Old South: The Failure of Agricultural Reform* (1988); Betty L. Mitchell, *Edmund Ruffin: A Biography* (1981); William K. Scarborough, ed., *Diary of Edmund Ruffin*, 3 vols. (1972–89).

St. Jude Children's Research Hospital

St. Jude Children's Research Hospital in Memphis, Tenn., was founded by the late entertainer Danny Thomas. The hospital, which opened in 1962, is the realization of a promise Thomas made as a struggling actor years before to St. Jude Thaddeus, the patron saint of hopeless causes. Thomas had vowed, "Show me my way in life, and I will build you a shrine."

In the 1950s Thomas consulted friends about what form this vow might take, and he decided to build a children's hospital, devoted to finding cures for catastrophic childhood diseases, in Memphis. The city was centrally located and had a large medical community and an established medical school at the University of Tennessee. Supporters came together across racial, ethnic, religious, regional, and national divisions to alleviate the suffering of children from leukemia, sickle cell anemia, and other diseases. Support for St. Jude also helped build powerful coalitions between the business and research economies and undermined local segregation practices. Construction of the St. Jude campus began in the 1950s and was completed in the 1960s when Memphis hospitals and other public accommodations were racially segregated and city residents were ambivalent about the treatment of indigent patients coming to Memphis from the rural countryside. But St. Jude's commitment to medical treatment without regard for race, religion, income, or residence positively influenced practices in the city of Memphis and in other southern cities.

Thomas and his wife, Rose Marie, traveled all over the United States raising money to build the hospital. He then turned to his fellow Americans of Arabic-speaking heritage, and in 1957 they founded the American Lebanese Syrian Associated Charities (ALSAC) as a way to show thanks to the United States for providing freedom and opportunity to their parents. ALSAC became the fundraising organization of St. Jude Children's Research Hospital and exists today solely to raise the funds necessary to operate and maintain the hospital. The daily operating cost of St. Jude is nearly $1.4 million dollars.

The mission of St. Jude is to find cures for children with cancer and other catastrophic diseases through research and treatment. St. Jude treats upward of 250 patients per day—on average, more than 5,400 children annually—most on an outpatient basis. The hospital maintains 78 beds for patients who require hospitalization. St. Jude is the only pediatric cancer research center where families never pay for treatment not covered by insurance. No child is ever denied treatment because of the family's inability to pay. Patients come from all 50 states and from around the world.

Patients and a family member are provided airfare to and from Memphis and are housed free of charge.

The current research at St. Jude includes work in gene therapy, chemotherapy, radiation treatment, hereditary and blood diseases, pediatric AIDS, and the psychological effects of catastrophic illnesses. St. Jude has developed protocols that have helped push overall survival rates for childhood cancers from less than 20 percent when the hospital opened in 1962 to more than 70 percent. It has one of the largest pediatric sickle cell disease programs in the country, which built upon the early efforts of researcher and pathologist Lemuel Diggs at the University of Tennessee in Memphis. The scientific discoveries are shared freely with medical communities throughout the world, and teams of doctors come to St. Jude to study and learn protocols, which they can take back home. St. Jude is the only pediatric cancer center to be designated as a comprehensive cancer center by the National Cancer Institute.

The location of St. Jude has helped create a vibrant neighborhood in a previously marginalized area of low-income housing projects and has become part of the historic urban fabric of Memphis. Recent construction more than doubled the size of the campus, creating jobs and providing a substantial employment and residential base in downtown and revitalizing the local urban infrastructure. St. Jude now has more than 20 buildings and 2.5 million square feet of research, clinical, and administrative space dedicated to finding cures and saving children.

The campus is close to downtown, one of the most diverse areas of Memphis, which includes governmental, medical, cultural, commercial, sports, and entertainment services and agencies. The hospital campus is a close neighbor of Uptown, a resurgent historic 100-block neighborhood that is a private-public revitalization by downtown development pioneers Henry Turley and Jack Belz along with the city of Memphis.

St. Jude Children's Research Hospital was the first hospital in Memphis to be integrated. Dr. Diggs, one of the original members of the St. Jude Board of Governors, wrote an impassioned letter to ALSAC CEO Mike Tamer on the issue of integration. Dr. Diggs detailed how one of his lab technicians had just lost his son to leukemia and how in addition to that terrible loss, the family also had $20,000 in medical bills. "The petty matters of race pale into unimportance in the face of catastrophes of this type," Diggs wrote.

In addition to being the first integrated hospital in Memphis, St. Jude played a key role in the integration of Memphis area hotels. When the hotel that St. Jude had contracted with to house patients and their families refused to provide rooms for African American families, then–St. Jude Director Dr. Donald Pinkel issued an ultimatum: If the hotel refused to accept African American children being treated at St. Jude, their services would not be used for any patient being treated at St. Jude.

PALLAS PIDGEON
Memphis, Tennessee

Keith Wailoo, *Dying in the City of the Blues: Sickle Cell Anemia and the Politics of Race and Health* (2001); Wanda Rushing, *Memphis and the Paradox of Place: Globalization in the American South* (2009).

Savannah River Site

When it was built in the early 1950s, the Savannah River Site was hailed as an engineering marvel on par with the construction of the Panama Canal. Containing five nuclear reactors, other large-scale nuclear and chemical facilities, high-tech research centers, multiple waste treatment sites, and a host of other support buildings, the entire site is spread over 310 square miles of mostly wooded land owned by the U.S. government on the western border of South Carolina, approximately 30 miles southeast of Augusta, Ga. Construction began on the site in 1950 when the Atomic Energy Commission contracted with E. I. du Pont de Nemours and Company to build and run a major facility that could produce the key materials for America's nuclear weapons on an industrial scale, particularly tritium and plutonium-239. As the U.S. government responded to the growing challenges of the Cold War and the arms race with the Soviet Union, this new plant formed a crucial part of America's nuclear defense network, as well as ongoing national research programs in nuclear, chemical, and environmental processes. Perhaps not quite as famous as sites in Oak Ridge, Tenn., or Los Alamos, N.Mex. (though it plays an important role in the 2002 film *The Sum of All Fears*), the Savannah River Site was in many ways the nucleus of America's atomic age, and it continues to manage the legacies of that era while promoting new forms of scientific research and technological innovation.

Among the numerous variables that helped determine the site's location, it was significant that the larger region of the Central Savannah River Area had a long history of industrial development dating back as far as the 1830s and 1840s, when local entrepreneurs brought some of the first textile and flour mills to the South. Yet it was inevitable that the site's impact on the surrounding area would be greater than any other industrial force preceding it. During construction, the towns of Ellenton and Dunbarton, as well as smaller hamlets, were completely evacuated and dismantled; many of Ellenton's original buildings were pulled off their foundations and moved to what is now New Ellenton, just outside the site's northern boundary. The site also brought significant change to nearby Aiken, S.C., where scientists, professionals, and other skilled workers relocated from all over the country, mixing with the long-established, southern white and African American residents as well as the wealthy, mostly northern socialites who regularly spent their winters there. As a social critic wrote at the time, "It [was] as if Scarlett O'Hara had come home from the ball, wriggled out of her satin gown, and put on a space suit."

The site is still managed and run through agreements between the Department of Energy and different subcontractors. And though none of the site's original reactors still operate as they did, the mission of the Savannah River Site has evolved as global politics have shifted. In addition to continuing to support the stabilization of existing nuclear weapons, one of the site's main operations now involves waste management, particularly with the Defense Waste Processing Facility, which converts radioactive waste from America's defense program into glass for long-term storage. Other parts of the site, including the Savannah River National Laboratory and the Savannah River Ecology Lab, conduct experimental research for a wide range of applications, such as fuel production, nonproliferation, and environmental cleanup and management. Although the Cold War may be technically over, the Savannah River Site remains a vital component of America's national security program and a center for scientific and technological advancement whose influence extends far beyond the region that surrounds it.

MICHAEL P. BIBLER
University of Manchester

Dorothy Kilgallin, *Good Housekeeping* (May 1953); Daniel Lang, *New Yorker* (7 July 1951); Mary Beth Reed, Mark Swanson, Steve Gaither, J. W. Joseph, and William Henry, *Savannah River Site at 50: Proceedings of the South Carolina Historical Association* (1994).

Scopes Trial

The Scopes antievolution trial took place in the small town of Dayton, Tenn., 10–21 July 1925. The participation of William Jennings Bryan, the thrice-defeated Democratic presidential candidate, and Clarence Darrow, the celebrated criminal defense attorney, ensured that the trial would gain national attention, and the Scopes Trial remains the best-known clash between evolution's defenders and activists who seek to ban Charles Darwin's theory from the public schools.

Dayton became involved in the controversy almost by happenstance. The South had seen scattered skirmishes over teaching evolution ever since the publication of Darwin's *On the Origin of Species* in 1859, but these fights were confined to the college level. Public schools seldom taught the subject, and only a small percentage of youths stayed in school until the later years of secondary education, when evolution occasionally made its appearance. The issue became prominent only when northern fundamentalists, who proclaimed that evolution violated their belief in the Bible as the literal, inerrant word of God, turned their attention to the South. Under prodding from Bryan, who led the crusade, Kentucky lawmakers proposed the first statewide antievolution bill, but the legislature narrowly rejected it. However, in 1925 Tennessee passed the "Butler Act," making it a crime to teach evolution because it "denies the story of the Divine creation of man as taught in the Bible." Governor Austin Peay, who needed allies in his far-reaching program to modernize Tennessee's schools, signed the act into law.

The newly founded American Civil Liberties Union (ACLU) advertised for a teacher willing to create a test case for the law. Under prodding from Dayton "boosters" who saw an opportunity to publicize their small town, a young high

William Jennings Bryan (seated at left) being interrogated by Clarence Seward Darrow, during the trial of The State of Tennessee v. John Thomas Scopes, 20 July 1925. *That Monday afternoon, because of the extreme heat, Judge Raulston moved court proceedings outdoors. The session was held on a platform that had been erected at the front of the Rhea County Courthouse to accommodate ministers who wanted to preach during the time of the trial. Defense lawyers for Scopes (John R. Neal, Arthur Garfield Hays, and Dudley Field Malone) are visible seated to the extreme right. One of the men at left, with his back to the photographer, appears to be Scopes. The court reporters are seated at the table. (Photo courtesy Smithsonian Institution, Washington, D.C.)*

school teacher named John Thomas Scopes agreed to be arrested for the misdemeanor of violating the Butler Law. The boosters' hopes were fulfilled when Bryan volunteered for the prosecution and Darrow joined the ACLU in Scopes's defense. The "trial of the century," as it quickly became known, attracted reporters from throughout the nation and was broadcast over the first radio network, which a Chicago station had set up expressly for the trial.

Although the trial was celebrated as a duel between science and religion, it raised numerous other issues. At a time when Tennessee's school system was rapidly expanding, the trial pressed the question of whether education should be controlled by majority rule, as through the legislature, or by educational experts, such as the scientists who testified at the trial. The sides also raised arguments over the separation of church and state, although the Supreme Court did not yet apply the First Amendment's Establishment Clause to the states. The best-known moment of the trial came on the seventh day, when Darrow cross-examined Bryan as a hostile expert witness and persuaded him to admit that the "days" of creation in Genesis might actually have lasted thousands of years—a common fundamentalist position for the time, but one that was damning for the purposes of the trial. Nevertheless, the jury found

Scopes guilty. The defense planned to appeal the law up to the Supreme Court, but the Tennessee Supreme Court reversed the conviction on a technicality, thus ending the legal debate.

The South saw a flurry of antievolution activity in the next three years, including Arkansas's 1928 passage of the only antievolution measure ever put to popular referendum. By that time, the antievolution movement had reached its geographic limit and made little headway in the North. Even in the South, legislatures passed no further major antievolution legislation, but less formal local pressure, as well as teachers' own beliefs, nearly quashed the teaching of evolution until national science curricula began to be adopted in the 1960s. Tennessee repealed the Butler Law in 1967.

The Scopes Trial set no legal precedents. Although it was partly a consequence of educational progress in Tennessee, it firmly affixed the label of backwardness to southern education. Decades later, the play *Inherit the Wind* and the film derived from it were loosely based on the Scopes Trial, though their message was aimed more at the anticommunist hysteria of the 1950s than the fundamentalism of the 1920s. The South remains the region most sympathetic to antievolutionism.

JEFF MORAN
University of Kansas

Edward J. Larson, *Summer for the Gods: The Scopes Trial and America's Continuing Debate Over Science and Religion* (1997); George M. Marsden, *Fundamentalism and American Culture: The Shaping of Twentieth-Century Evangelicalism, 1870–1925* (1982); Jeffrey P. Moran, *The Scopes Trial: A Brief History with Documents* (2002); Willard B. Gatewood, *Preachers, Pedagogues, and Politicians: The Evolution Controversy in North Carolina, 1920–1927* (1966).

Sickle Cell Anemia

Hemoglobin is the iron-rich protein in red blood cells that carries oxygen from the lungs to the rest of the body. People who inherit an abnormal hemoglobin gene (hemoglobin S) from one parent and a normal hemoglobin gene from the other parent have a condition called sickle cell trait. Though their blood contains some sickle-shaped cells along with regular disk-shaped red blood cells, they lead generally normal lives with few overt symptoms. But people who inherit two hemoglobin S genes—one from each parent—are born with sickle cell anemia, or sickle cell disease (SCD), which gives them a higher proportion of sickle-shaped red blood cells. This lifelong illness is characterized by chronic fatigue and jaundice, occasional episodes of severe pain, and shortened life expectancy. (It takes two parents, each with sickle cell trait, to produce an infant with sickle cell disease, but not all of their children will automatically have SCD.)

Because sickle cell trait in both parents can lead to debilitating SCD in a child, one might expect natural selection to remove this harmful trait. But it has persisted for centuries in equatorial and semitropical areas, because the presence of some sickle cells also provides partial protection, especially in early childhood, against the devastating effects of mosquito-borne malaria.

(Scientists refer to these rare instances of offsetting negative and positive effects in genes as "balanced polymorphisms.") Global maps of historical malaria distribution coincide impressively with maps showing the distribution of both sickle cell trait and SCD.

Most persons deported from sub-Saharan Africa to the New World through the Atlantic slave trade came from regions where malaria was prevalent. Therefore, sickle cell trait became commonplace in the American South during colonial times. Moreover, malaria was one of many Old World diseases brought to the Americas during the so-called Columbian Exchange that began in 1492. So sickle cell trait, with its positive and negative effects, persisted in the American South. It gave an advantage to many African newcomers over Europeans and Native Americans in coping with the malarial fevers that recurred annually in the warm coastal South during mosquito season, but it also impaired their health in other ways and caused SCD in subsequent African American generations. Sufferers can experience joint pain, leg ulcers, breathlessness, and swelling of the hands and feet. Despite these varied symptoms, however, the actual disease remained unrecognized, owing largely to medical inattention and racial bias.

Only in the past century has research finally identified this serious sickness and managed to explain its genetic cause and uneven distribution. Scientific awareness began in 1910, when a Chicago doctor described the strange case of a dental student from Granada who had been treated over several years for "muscular rheumatism" and "bilious attacks." The patient's blood contained "peculiar elongated and sickle-shaped" cells that contrasted with the round disk form of normal red blood cells. A dozen years later, a medical resident at Johns Hopkins named this condition "sickle-cell anæmia," but it was not until 1949 that Dr. Linus Pauling and colleagues showed that the illness resulted from an abnormality in the hemoglobin molecule. Their paper, "Sickle Cell Anemia, a Molecular Disease," became a landmark in the history of molecular biology, because it linked a genetic disease to a mutation of a specific protein for the first time.

The genetic mutation involved hemoglobin, the iron-rich protein in red blood cells that carries oxygen from the lungs to the rest of the body. Normal red blood cells, which look like doughnuts without holes in the center, flow easily through the blood vessels and live roughly 120 days. In contrast, abnormal red cells, containing hemoglobin S, develop a sickle shape. These distorted cells last scarcely 20 days, so they carry less oxygen. Also, their crescent shape can obstruct blood vessels and block the circulation, causing pain, infection, or organ damage. For example, a sickle cell crisis can harm the spleen, which normally generates new red blood cells and processes old ones. Such repeated "splenic infarctions" can end the spleen's function, the equivalent of surgically removing the organ. (This may explain why a note in the *Southern Journal of Medical Pharmacology* in 1846, describing the autopsy of a black slave, noted the absence of a spleen.)

With the gradual eradication of malaria across the South during the 20th century, sickle cell trait became more a liability than an asset, and its presence continues to diminish over generations. At present, nearly 3 million Americans have sickle cell trait. Screening shows its presence in roughly 1 in 12 African Americans, but people with Mediterranean, Spanish, or Asiatic Indian ancestry are also affected. As a consequence, sickle cell disease currently affects more than 90,000 Americans. SCD occurs in about 1 out of every 500 African American births and in a smaller proportion of Latino American births. Through genetic screening, doctors can now alert couples likely to pass on the disease, diagnose its early presence, manage debilitating symptoms, and limit the frequency of crises. As yet, however, no widespread cure exists.

PETER H. WOOD
Duke University

A. C. Allison, *Cold Spring Harbor Symposium on Quantitative Biology* (1964); Center for Disease Control and Prevention, www.cdc.gov/Features/SickleCell (22 May 2011); David J. Wetherall, *Current Molecular Medicine* (November 2008); Michael Aidoo et al., *Lancet* (13 April 2002); *ScienceDaily* (2 November 2010); Keith Wailoo, *Dying in the City of the Blues: Sickle Cell Anemia and the Politics of Race and Health* (2001).

Sims, J. Marion

(1813–1883) PHYSICIAN.

James Marion Sims was born near Hanging Rock Creek, Lancaster District, S.C., on 25 January 1813 and named Marion in honor of Gen. Francis Marion, the "Swamp Fox" of the Revolutionary War era. He pursued a bachelor's degree at South Carolina College and graduated in 1832. His father, John Sims, strongly disapproved of a decision to study medicine but arranged for an apprenticeship under preceptor Dr. Churchill Jones prior to his son's enrollment at the Medical College of the State of South Carolina in Charleston in 1833. Following graduation and another short spell as an apprentice, Sims attended Jefferson Medical College, Philadelphia, in May 1835, where he remembered being inspired by Dr. George McClellan's daring surgical operations. Returning to Lancaster District, Sims married childhood sweetheart Eliza Theresa Jones in 1836. The couple had at least five children. However, after a disastrous start to his career, with the death of two infant patients, the Sims family migrated to Alabama, where he eventually established a successful practice in Montgomery and gained a reputation for bold surgical interventions in cases of cross-eye, clubfoot, harelip, and tumors of the jaw. Sims's career here became deeply intertwined with the institution of chattel slavery, as he became a slave owner and physician to a wide range of slaveholding clientele.

Between 1844 and 1849, Sims performed numerous surgeries on enslaved people, many conducted in a purpose-built private slave hospital. These included experiments involving enslaved infants for the surgical relief of neonatal tetanus and surgical trials on enslaved women to repair vesicovaginal fistula (a complication of childbirth). Because of racial prejudice and their status as property, enslaved people often found themselves exploited for such medical experiments in the antebellum South. In his autobiography, *The Story of My Life*, Sims boasted that in this most "memorable era" of his life "there was never a time that I could not, at any day, have had a subject for operation." One young enslaved woman, Anarcha Wescott, herself alone underwent more than 30 operations without anesthesia, as Sims used the opportunity to design various gynecological instruments—the most notable of which were his speculum and wire sutures—and new operative procedures, including the eponymous genu-pectoral or Sims position.

Perfecting his technique in repeated experiments on enslaved patients, Sims made his breakthroughs known to the wider medical profession in an article published in the *American Journal of Medical Sciences* in 1853. That same year Sims and his family sold their slaves and other assets, moving to New York and setting up home on Madison Avenue. Repeating a key element of his professional self-making in Montgomery, Sims established a charity hospital under his control for the instruction of fellow physicians. The Woman's Hospital opened on 4 May 1855 and was later chartered as the Woman's Hospital of the State of New York.

Sims spent the early 1860s avoiding the Civil War and visiting major European medical centers, displaying his distinctive method of operation to many interested observers. He also acquired an elite referral practice that included members of the aristocracy, such as the Duchess of Hamilton, Empress Eugénie, and the Empress of Austria. He eventually returned to New York in 1868 but maintained consultation rooms in Paris and London. Sims died in New York on 13 November 1883 of coronary artery disease and was buried in Greenwood Cemetery.

Toward the end of his life, Sims took stock of his social and professional standing. He calculated himself to be the second wealthiest of all American doctors and had achieved the honor of his profession. He was made president of the American Medical Association in 1875 and the American Gynecological Society in 1880, receiving medals, decorations, and awards from the governments of France, Italy, Belgium, Spain, Portugal, and Germany. For most of the 20th century, Sims enjoyed a posthumous reputation as the "Father of Gynecology." Physician-authored medical histories and monuments celebrated Sims as a medical innovator, and he is said to have brought relief to generations of countless numbers of women suffering accidents of childbirth. With the rise of women's history, the social history of medicine (now importantly incorporating the patient's perspective), and African American history in the early 1970s, Sims's career received a thorough reconfiguration from the perspective of his slave, pauper, and female patients. Although traditional medical historians continued to emphasize Sims's technical achievements, by the late 1990s the overwhelming popular and academic consensus became that

Sims was self-serving, unethical, and deeply prejudiced in many of his surgical experiments.

STEPHEN C. KENNY
University of Liverpool

Seale Harris, *Woman's Surgeon: The Life Story of J. Marion Sims* (1950); Deborah Kuhn McGregor, *From Midwives to Medicine: The Birth of American Gynecology* (1998), *Sexual Surgery and the Origins of Gynecology: J. Marion Sims, His Hospital, and His Patients* (1990); Jeffrey S. Sartin, *Southern Medical Journal* (May 2004); James Marion Sims, *The Story of My Life* (1884).

Slave Hospitals

Hospitals specializing in the care of enslaved patients date back to the earliest years of black presence in the New World and served various roles and white interest groups as the institution of slavery adapted and changed over time. Different types of hospitals were provided for enslaved patients: plantation hospitals, medical school hospitals, private infirmaries operated by individual doctors or physician partnerships, and commercial hospitals with strong links to the domestic slave trade.

In the era of the transatlantic slave trade, one primary function of slave hospitals was to act as pesthouses, lazarettos, and quarantine stations: places to contain, segregate, inspect, and season newly imported Africans and ready them for their next appearance at market. The threat of contagions introduced by slave-trading vessels, and an outbreak of yellow fever in 1706, led Carolina's colonial governors to introduce quarantine regulations in 1707 and establish a pesthouse on Sullivan's Island. Hospitals also became a standard feature of large Lowcountry plantation estates, with slave owners responding to the need for conveniently located facilities devoted to slave health and maintenance, as well as to the constant menace of epidemic fevers among growing captive populations in a malarial region.

Although many large plantation complexes included buildings described as slave hospitals, these were often little more than a cabin used to isolate the afflicted, or were simply somewhere to keep close watch over the sick and ensure they were not feigning illness or planning to run away. Even where dedicated slave hospital facilities did exist, they were often unhygienic and poorly managed. In a journal account of her residence at Butler Island plantation in Georgia, Frances Kemble remembered sick slave women lying on the floor of the large infirmary in "tattered and filthy blankets."

The official closure of the transatlantic slave trade to North America in 1808 and the expansion of a vigorous domestic trade in human beings probably acted as a stimulus to the development of slave hospitals. As the medical profession developed, doctors also recognized that large slave holdings and a wealthy planter class presented opportunities to establish slave hospital facilities and lucrative practices. By the 1840s and 1850s, dozens of slave infirmaries were operating in South Carolina, Georgia, Mississippi, Alabama, Tennessee, Virginia, and Louisiana. Often, these slave hospitals where located in major towns and cities with large slave populations and multiple slave-trading depots and auction sites servicing the domestic chattel market—slave-trading centers such as Columbia, Charleston, Augusta, Montgomery, Memphis, New Orleans, Mobile, and Savannah. But slave hospitals could also be found in more remote rural locations, such as Blackville, S.C., and Kosciusko, Miss.—functioning as both hospitals and way stations for the transportation of chattel slaves.

Most major medical colleges across the South adopted hospital facilities for the treatment of slaves during the antebellum era, as physicians realized the potential of utilizing slave patients in medical education and clinical research. The Medical College of South Carolina's slave hospital became a unique selling point for the institution and a clinical medical laboratory, providing physicians with a constant supply of research subjects. Similar facilities were in operation at Richmond's College Infirmary of Hampden-Sydney Medical Department and at the Medical College of Georgia in Augusta.

In the early 1840s physician J. Marion Sims erected a hospital behind his office on Perry Street in Montgomery, Ala., primarily for the care of "negro surgical cases." Here he performed experimental surgical operations on enslaved men, women, and infants. Charleston surgeon Dr. Julian John Chisolm also published reports of dozens of operations performed on slave patients at his private infirmaries and the college hospitals where he was a professor of surgery.

Many experimental and commercial southern slave hospitals played important roles in restoring, ensuring, and enhancing the value of slave property. New Orleans's Touro Infirmary received slave patients from a wide range of clients and businesses but also numerous patients whose treatment was paid for by slave traders. Notorious slave dealer Bernard Kendig was one of more than a dozen slave-trading individuals who sent patients to Touro Infirmary between 1855 and 1860. Slave patients were admitted to Touro with fevers, physical exhaustion, and malnutrition, having being forcibly transported enormous distances by road, rail, river, and ocean, and also with skin and lung diseases contracted in the cramped and filthy pens of large slave traders. Touro's house physician, Dr. Bensadon, provided not only treatment to slave patients but also certificates of soundness to traders to facilitate and legitimate slave sales. Across the South, the slave infirmary functioned as a key part of the network of slave markets, exchanges, and auction houses, serving the business needs of significant numbers of slave traders and speculators who sought to reassure prospective buyers of slaves that they need not worry about the possibility of receiving damaged goods. By the late antebellum era, slavery and medicine had developed mutual interests to such an extent that medical professionals were providing dedicated hospital facilities to accommodate and service the needs of the region's lucrative trade and investment in slave bodies.

STEPHEN C. KENNY
University of Liverpool

Katherine K. Bankole, *Enslavement and Medical Practices in Antebellum Louisiana*

(1998); Frances Anne Kemble, *Journal of a Residence on a Georgian Plantation in 1838–1839* (1863); Todd L. Savitt, *Medicine and Slavery: The Diseases and Health Care of Blacks in Antebellum Virginia* (1981).

Tuskegee Syphilis Study (United States Public Health Service Syphilis Study)

The United States Public Health Service Study of Untreated Syphilis (USPHSS) in the African American male in Macon County, Ala., is the original name of the longest nontherapeutic study conducted in the United States. This study is more popularly known as the Tuskegee Syphilis Study because the United States Public Health Service originally conducted it with cooperation from Tuskegee Institute (aka Tuskegee University). The study continued from 1932 to 1972. After its founding in 1881, Tuskegee Institute worked with members of the black community in Macon County and the surrounding Black Belt counties to improve the well-being and standard of living for black people who resided in those counties. The institute designed several extension programs centered on education, farming, animal husbandry, housing improvements, nutrition, and health. Before the opening of the Macon County Public Health Department in 1946, all public health functions by the state of Alabama for black people in Macon County and several surrounding counties were conducted at the John A. Andrews Memorial Hospital, a private hospital operated by Tuskegee Institute. The proposed research study of syphilis was originally sponsored by the Rosenwald Health Fund, as a diagnostic and treatment program in six southern states, including Alabama, to address the national concerns about the syphilis epidemic that was expanding throughout the country.

The stock market crash forced the Rosenwald Fund to reevaluate its program funding and the original study was dropped. In an effort to salvage the program, the U.S. Public Health Service proposed a more limited study of untreated syphilis, as a comparison study, to the retrospective study conducted in Oslo, Norway, in 1925. The USPHSS began with approximately 600 men, 400 diagnosed with syphilis and 200 without the disease. When the study began, arsenic and mercury were used to treat those people infected with the disease. The men with syphilis were not treated. A concerted effort by the U.S. Public Health Service, the Alabama State Health Department, and local physicians was coordinated to assure that even those men without the disease, who were later presented with syphilis, were not dropped, but were switched and included with the infected population.

The USPHSS did not stop in 1943, even when the U.S. Public Health Service determined that penicillin was the most effective method for treating venereal diseases, particularly syphilis and gonorrhea. Even the Nuremberg Trials and the declaration against human experimentation without consent did not stop the USPHSS. Nor did the scientific community raise any concerns

Dr. Walter Edmondson taking a blood test from an unidentified subject (National Archive, Atlanta, Ga.)

about unethical behavior by the research team or plausible bioethical violations by the U.S. Public Health Service or the Centers for Disease Control (CDC) until the 1950s. Over the 40-year period (1932–72), several articles were published in the scientific literature about findings from the study. Additionally, information gathered from the study was included in the curriculum of many medical schools throughout the country.

By the late 1960s and early 1970s several employees at the CDC expressed grave concerns regarding the study. Nonetheless, the study continued until 1972, when an Associated Press article, written by Jean Heller, was published in the *New York Times*. The revelation resulted in a class-action suit filed in July 1973 by noted civil rights attorney Fred D. Gray and, among other study participants, Mr. Charlie W. Pollard. A settlement of approximately $10 million was reached in 1974.

On 16 May 1997 President William Jefferson Clinton apologized to the surviving men and their families, the Tuskegee community, and the African American population on behalf of the country for the study. At a White House ceremony in Washington D.C., President Clinton said the government did something that was "wrong, gravely and morally wrong." Five of the remaining seven study survivors, including Charlie Pollard, were present. As a part of the apology, President Clinton mandated that the Department of Health and Human Services support the establishment of the Tuskegee University National Center for Bioethics in Research and Health Care on the campus of Tuskegee University and a legacy museum be launched. In 1999 the Bioethics Center opened on the campus of Tuskegee University to conduct research, education, and community engagement activities related to bioethics, public health ethics, health disparities, and health equity. After more than a decade, the Bioethics Center continues its mission.

RUEBEN WARREN
Tuskegee University

Fred D. Grey, *The Tuskegee Syphilis Study* (1998); James H. Jones, *Bad Blood: The Tuskegee Syphilis Experiment* (1993); Ralph V. Katz et al., *American Journal of Public Health* (June 2008); Susan M. Reverby, *Examining Tuskegee: The Infamous Syphilis Study and Its Legacy* (2009), ed., *Tuskegee's Truths: Rethinking the Tuskegee Syphilis Study* (2000).

Whitfield (Mississippi State Hospital)

Whitfield is the colloquial name for the Mississippi State Hospital, Mississippi's primary public mental institution and hospital, which dates to 1848. The name "Whitfield" is derived from the post office and railroad station located at the hospital. Situated 10 miles southeast of Jackson, the post office was named in memory of Gov. Henry L. Whitfield, a Rankin County native who served as Mississippi's chief executive from 1924 to 1927, in office when the state legislature voted to relocate the Mississippi State Insane Hospital, as it was then known, from Jackson to its current Rankin County site. The history of Whitfield is a case study of how southern states developed and evolved institutions to deal with the mentally ill.

The modern treatment of mental illness in Mississippi dates to antebellum times. Before the late 18th century, the mentally ill, then termed "lunatics" and "idiots," often were not considered worthy of public concern. They frequently wandered the streets or were kept locked up by families at home, with the violently psychotic sometimes chained to the floors of jails. Early mental institutions developed not to treat the afflicted but rather to confine them away from the general public. By the late 18th and early 19th centuries, physicians such as Philippe Pinel and Benjamin Rush encouraged a more scientific and humane approach that resulted in the creation of benevolent institutions and hospitals for the scientific treatment of the insane. In the early 1840s leaders in the state's medical community, especially Drs. William S. Langley, Edward Pickett, and Thomas J. Catchings, championed the idea of erecting such a hospital in Mississippi. In January 1846 Gov. Albert G. Brown proposed the erection of "an asylum for lunatics" and "a refuge for the insane" in the state. Two years later legislation appropriated $10,000 and a five-acre lot in the city of Jackson. An early superintendent later remarked of this limited appropriation that Mississippi's asylum was "born in debt" and spent most of its early history "begging and borrowing."

The asylum's commissioners soon purchased a tract of 140 acres of land two miles north of Jackson (which is now the present location of the University of Mississippi Medical Center), and on that site in 1848 work began on a large central building with two wings. Work on the project ceased when the scanty original funding was exhausted. In 1850 the commissioners requested $50,000 more to complete the project. Assisted by the support and presence of Dorothea L. Dix, a Boston schoolteacher nationally known as a mental health reformer, the commissioners crystallized public support and secured the necessary funding from the legislature. By 1851 the first buildings of the "Mississippi State Lunatic Asylum" were erected and a cornerstone placed, but it took four more years of construction and further state funding before the asylum opened its doors to patients in 1855.

The first superintendent, Dr. William Langley, had been among the first proponents of the asylum's establishment. By 1856, at the request of the asylum trustees, admission of slaves and free persons of color was initiated on a small scale. In its early years the Jackson asylum survived fires, tornadoes, yellow fever epidemics, and shifting Yazoo clay. The main building, with six marble columns and a classic front, crowned with a cupola, had wing after wing added on to its sides and sprawled out like a prehistoric bird on the outskirts of early Jackson. For generations, it provided care for thousands of Mississippi's mentally ill.

During the Civil War, both Union and Confederate troops occupied the grounds, digging extensive earthen fortifications and rifle pits. The Union forces camped around the hospital, stripping it of its cows, hogs, and garden produce. During this federal occupation in July 1863, there was considerable skirmishing between the armies, and Confederate forces fired several shells upon the asylum, with minor injury to the buildings. Superintendent Robert Kells, with the support of Gov. Charles Clark, kept the asylum open throughout the war. The poorly funded institution continued to operate after the war. In 1870 Gov. James Alcorn appointed Dr. William Compton as superintendent. A nationally recognized mental health physician, Compton utilized his great political skill to secure gubernatorial and legislative support. He also embarked on efforts to modernize the medical treatment of the insane at the facility and to double its capacity. He requested improved lodging for the "lunatics of color" to equal that provided for whites, while acknowledging, with the support of Governor Alcorn, the need for segregation by both race and sex.

Overcrowding continually plagued the institution. To relieve this, a second hospital, East Mississippi Insane Asylum (now East Mississippi State Hospital), was established in Meridian on 8 March 1882 to help treat the state's mentally ill. By 1890, more room was still needed at the Jackson asylum. After discussions of opening an institution in the Delta, the legislature authorized in 1890 and 1900 the building of additional annexes at Jackson for increasing numbers of black patients.

In January 1900 the lunatic asylum changed its name to the Mississippi State Insane Hospital. In the first two decades of the 20th century, the old institution continued to deteriorate physically as its census swelled to 1,350 beds. In 1918 Dr. Charles Mitchell was appointed as superintendent. During his 21-year administration, he ushered in many psychiatric advances and advocated relocating the hospital to create a more modern campus. By 1926 the Jackson hospital reached a census of 2,000 patients, and the grounds totaled more than 1,300 acres, which were farmed by the patients. In this same year, the legislature appropriated $2.5 million for a new hospital to be located on 3,333 acres of state-owned land "near Howell" in Rankin County. The 1926 legislation also dropped "Insane" from the hospital's title, and it became simply "Mississippi State Hospital." However, because of a significant drop in state income during the Depression and ongoing political squabbles, the hospital did not open at the new site until 4 March 1935, at a total cost of $5 million.

Relocating the patients from the institution on North State Street in Jackson to Whitfield was conducted in March 1935 and took only nine days. The Whitfield campus was a more isolated and pastoral environment, based on a modern prototype very different from the interconnected wards and annexes of the old asylum. Highly regarded architect N. W. Overstreet planned the main campus, which covered 350 acres and consisted of more than 75 red brick buildings built in colonial design with white columns and trim. The original plan included two segregated campuses. The western side of the main campus was for "colored" residents and the eastern side for whites. Whitfield remained segregated racially until the passage of the Civil Rights Act of 1964, when administrator Dr. William "Jake" L. Jaquith desegregated the campus uneventfully.

The new hospital's capacity was 3,500 patients. It had its own electric power plant, ice plant, laundry, bakery, store, telephone service, canning plant, post office, sawmill, carpentry shop, railway station, dairy, farms, orchards, water works, and nursery. By 1955 the institution at Whitfield had grown to 4,000 patients, with more than 800 employees. However, after increasing legal challenges to the confinement of the mentally ill, the patient population significantly decreased to 2,600 by 1978 and to 1,600 by 1983. In the early 1980s there were even legislative discussions to close the institution. As community-based psychiatric programs have emerged, plans to downsize the institution have been embraced by mental health authorities and the legislature, with several of the buildings housing patients closing. Despite this shift in mental health priorities, in 2009 the hospital contained 915 licensed psychiatric beds and 418 licensed nursing home beds, while also providing a variety of community service programs.

In February 2000 the Mississippi State Hospital Museum opened on the Whitfield campus in "Building 23," the original white male receiving building constructed in 1929. The museum offers a concise historical overview of the treatment of mental illness in the state, centering on the critical role played by the Mississippi State Hospital. Original hydrotherapy rooms, needle spray showers, and a fever box are included among the museum's exhibits.

LUCIUS LAMPTON
Journal of the Mississippi State Medical Association

Annual Reports of the Board of Trustees and Superintendent of the Mississippi State Lunatic Asylum for the Year 1870–1877, various publishers; Lucius Lampton, *Journal of the Mississippi State Medical Association* (April 2000, January 2003); William D. McCain, *The Story of Jackson* (1953).

Yellow Fever

"To no other great nation of the earth is yellow fever so calamitous as to the United States of America." That was the conclusion of the board of experts authorized by the U.S. Congress in 1879 to investigate the worst yellow fever epidemic in U.S. history—an epidemic that took 20,000 lives in the Mississippi Valley in one summer.

Yellow fever, Yellow Jack, and the Saffron Scourge were all monikers for a virus that made its way from West Africa to this hemisphere through slave trading in the 17th and 18th centuries. The virus caused high fever, delirium, organ failure, external bleeding through the eyes, nose, and gums, and internal bleeding leading to the telltale "black vomit." The "yellow" describes the jaundice in final stages of liver failure. How this terrible disease spread from person to person and from one town to the next remained a medical mystery for centuries.

Yellow fever had a greater effect on southern culture than any other disease, not because of the gruesome symptoms or high mortality, but because it touched every corner of southern life. Originally a disease of the northern port towns, including Boston, New York, and Philadelphia, yellow fever spread south along railroads and Mississippi River traffic. By the 1840s yellow fever and malaria were almost yearly occurrences in cities like New Orleans, becoming a mark of distinction between "insiders" and "outsiders" in the South—those who were acclimated to the fevers and those who were not.

With time, however, the yellow fever epidemics grew more severe and more fatal. A combination of expanding transportation, an influx of nonimmune immigrants moving south, and a series of El Niño cycles helped pave the way for the disastrous epidemics to come.

The yellow fever outbreaks of the 1860s and 1870s struck with the force of a natural disaster, overwhelming cities, local governments, resources, hospitals, and churches. Although anyone with means fled during epidemics, countless doctors, nurses, priests, and nuns stayed and died caring for feverish patients. Gravediggers proved scarce. Bodies were found in parks, on benches, and on sidewalks. Pine coffins and lime lined the streets. Wagons rattled by with calls to "Bring out your dead." Fever fires burned day and night. Refugee camps erupted through the countryside, from fairgrounds to chicken shacks. In Memphis, in 1878, death could be smelled three miles away.

Divisiveness came naturally during epidemics. Racism ignited as blacks were blamed for carrying the disease. The wealthy fled; the poor were left to die. Small towns refused urban citizens for fear of spreading fever. Commerce was shut down for months at a time as cities were quarantined. And barely a decade after the Civil War, the federal government had no recourse for managing state or regional disasters. Finally, in the wake of several yellow fever epidemics, the National Board of Health was created to provide relief and begin studying the disease.

As the 20th century approached, yellow fever epidemics in the South waned, although the cause of yellow fever remained a mystery. But yellow fever continued to threaten U.S. troops in tropical locations like Cuba and Panama. It was U.S. army surgeon Walter Reed, assigned to Cuba during the Spanish-American War, who finally demonstrated that mosquitoes carry and spread yellow fever. Clean water systems and better sanitation helped eliminate many of the mosquito's breeding grounds, and the mysterious saffron scourges ended. By the 1920s work was underway for the vaccine still in use today.

Yellow fever remains a medical success story, but it was one that came at great cost to scientists, physicians, citizens, army troops, and, perhaps most of all, the southern cities so ravaged by it.

MOLLY CALDWELL CROSBY
Memphis, Tennessee

Khaled J. Bloom, *The Mississippi Valley's Great Yellow Fever Epidemic of 1878* (1993); Jo Ann Carrigan, *The Saffron Scourge: A History of Yellow Fever in Louisiana, 1796–1905* (1994); Molly Caldwell Crosby, *The American Plague: The Untold Story of Yellow Fever, the Epidemic That Shaped Our History* (2006); Margaret Humphreys, *Yellow Fever and the South* (1992); J. M. Keating, *The Yellow Fever Epidemic of 1878 in Memphis, TN* (1879); Library of Congress, *Conclusion of Board of Experts* (1879); World Health Organization, *El Niño and Health*; Mississippi Valley Collection, University of Memphis; Philip S. Hench Walter Reed Yellow Fever Collection, University of Virginia; Simon R. Bruesch Collection, University of Tennessee Library of Medicine; Yellow Fever Collection, Memphis Library.

INDEX OF CONTRIBUTORS

INDEX

Page numbers in boldface refer to articles.

FSC
www.fsc.org
MIX
Paper from
responsible sources
FSC® C013483